Felipe López-Saucedo

Molecules that Changed the World

Table Sugar, Ethanol, Aspirin, and Ammonia

www.novapublishers.com

DOI: https://doi.org/10.52305/VYYI1586

Library of Congress Cataloging-in-Publication Data

ISBN: 979-8-89530-599-7 (Softcover)
ISBN: 979-8-89530-642-0 (eBook)

Published by Nova Science Publishers, Inc. † New York

Contents

Preface

Many of the substances we rely on daily go unnoticed and sometimes their availability is taken for granted. Yet, each one has a complex history of discovery, production, and global impact. This book explores four essential, yet often overlooked molecules: table sugar, ethanol, aspirin, and ammonia. It examines their chemical nature, background, and economic and social significance. Through a blend of science and storytelling, readers will uncover how these substances have shaped industries, medicine, and even our way of life. Finally, this book invites readers to imagine a world without these compounds, fostering a deeper appreciation for the power of material transformation.

Acknowledgment

The illustrations in this book were done by Belén Gómez-Lázaro.

Introduction

Carbon constitutes only 0.032% of the Earth's crust and ranks as the fifteenth most abundant element. Though scarce, carbon is the basis of life since it is present in amino acids that are fundamental for all living organisms. Its remarkable ability to bond with other elements enables the synthesis of unique and indispensable compounds and materials for modern life.

Throughout history, humans have exploited natural resources and modified their environment to enhance their quality of life, a feature that sets them apart from other forms of life. As a result, human progress has been relentless, from the invention of the wheel to the development of engines, satellites, and computers. Inventions in chemistry, like engineering, have followed an upward trajectory. Sugar, ethanol, aspirin, and ammonia exemplify humanity's mastery over elements to solve food and health issues.

The elucidation of quantum mechanics, the establishment of theories of chemical bonding, and the developments in computational chemistry have deepened our understanding of matter. At the same time, advanced instrumental techniques have enabled detailed analyses of molecular structures. These advances have driven progress in medicine, agriculture, and energy, yielding items we use day by day.

Looking ahead, the role of chemistry is clear: driving innovation while mitigating the ecological damage caused by past activities. This way, humanity must ensure that future generations inherit a habitable planet by improving processes and prioritizing sustainability for the sake of the common good.

Chapter 1

Table Sugar

"Hide not thy poison with such sugar'd words."
William Shakespeare

General Overview and History of Sugar

Sugars are simple carbohydrates of natural origin that are abundant in plants. In particular, table sugar is distinguished by its characteristic sweetness. Pure table sugar is also a processed food composed of a unique disaccharide. It has a high energy value, and the body metabolizes and assimilates it as glucose, either through acid hydrolysis or enzymatic action (Combes & Monsan, 1983). Glucose is metabolized by the body to produce adenosine triphosphate (ATP) and caloric energy (Boutros, 1962). ATP serves as an energy reservoir that cells use when they require it; thus, glucose is necessary in specific amounts for the body's proper functioning.

Table sugar or sucrose is also known as saccharose, white sugar, or just sugar, so these terms are used interchangeably throughout this chapter. The IUPAC (International Union of Pure and Applied Chemistry) name α-D-glucopyranosyl-β-D-fructofuranoside, is hardly used, even in scientific jargon. The etymology of the word saccharose traces back to the Latin word *saccharum*, which is derived from the Greek word *σάκχαρον*, pronounced 'sácjaron' or 'sucrose' to refer to table sugar. Meanwhile, the word "sugar" originates from the Hispanic Arabic *as-súkkar*, from the Arabic *sukkar*, from the Persian *shekar*, and possibly from the Sanskrit *śarkarā*, which means gravel.

Although sugar is present in all plants, only a few of them are used as sources of raw materials for commercial extraction and refining. The reasons for selecting only a few plants are the high percentage of the disaccharide and the period between cultivation and harvest. The main sources are sugarcane (*Saccharum spontaneum*, *Saccharum officinarum*, and other *Saccharum spp*), sugar beet (*Beta vulgaris*), sorghum (*Sorghum bicolor*), and coconut (*Cocos nucifera*). Sugarcane and coconut thrive in tropical climates, while sorghum and sugar beet are more adaptable to different climates, as they can be grown in tropical, subtropical, temperate, and semi-warm climates. In addition to these raw materials, there are geographical areas where local (native) species are similarly processed to extract sucrose (Canadian Sugar Institute, 2024).

Sugar has been in the kitchen pantry for centuries, but the history of sugar began much earlier with the domestication of plants in primitive crops. Sugarcane cultivation was known to the natives of New Guinea as early as 8000 BC, and later knowledge of the crop spread to Southeast Asia and India (Dinesh Babu et al., 2022). At the end of the Ancient Age, a novel method for obtaining sugar crystals was described in India. It marked the first advance in the compound's purification, making its storage and transport easier and facilitating its trade around the known world.

The modern history of sugar continues with the German chemist Andreas Sigismund Marggraf (1709–1782), who focused on isolating sucrose from beets using alcohol for extraction. His studies on sugar purification were further developed by one of his students, Franz Karl Achard (1753–1821), who applied engineering principles to design a modern process for large-scale sugar production (Harveson, 2024). Achard's advancements marked a second major milestone in the history of refined sugar, paving the way for what would eventually become the sugar industry.

The research on sucrose in the last quarter of the 19th century by the American scientist Bernhard Tollens (1841–1918) and the German Hermann Emil Fischer (1852–1919), helped to reveal the secrets of the reactivity of sugars, particularly on their connectivity and isomerism (Pérez, 1995). These studies formed part of the foundation of modern organic chemistry, decades before the American Gilbert Newton Lewis (1875–1946) developed his atomic model between 1902 and 1916 and before the invention of instrumental analysis and spectroscopy.

Research on sugars in the 19th and 20th centuries was so significant that five scientists were awarded the Nobel Prize in Chemistry in different years for their related contributions. Emil Fischer was awarded the second Nobel Prize in Chemistry in 1902 for his work on the synthesis of sugars and purines. Later, in 1929, the English biochemist Sir Arthur Harden (1865–1940) and the German scientist Hans Karl August Simon von Euler-Chelpin (1873–1964) jointly received the Nobel Prize for their investigations into the fermentation of sugars and the role of fermenting enzymes. In 1937, the English chemist Sir Walter Norman Haworth (1883–1950) was recognized for his research on the structure of carbohydrates and the synthesis of vitamin C. Finally, more than three decades later, in 1970, the Argentine physician and biochemist Luis Federico Leloir (1906–1987) received the Nobel Prize for discovering the role of sugar nucleotides in carbohydrate biosynthesis.

Data on Sugar

Sugar is a white or yellow crystalline powder known for its sweet taste and suitability for human consumption. It is highly water-soluble, dissolving at approximately 197 g per 100 ml at 20°C. Its chemical formula is $C_{12}H_{22}O_{11}$, with a density of 1.587 g/cm^3 and a molecular weight of 342.3 g/mol. Sugar has a decomposition point of 186°C; but the monosaccharides glucose and fructose, have melting points of 146°C and 103°C, respectively.

The safety data sheet (SDS) for sucrose (CAS 57-50-1) indicates that it is a stable compound, and according to the National Fire Protection Association (NFPA) 704 standard, it is considered to pose low danger; the color-coded diamond indicates health = 0 (blue), flammability = 1 (red), and reactivity = 0 (yellow), while under special risks (white), no indications are present. Like other carbohydrates, sucrose burns when exposed to flame and is incompatible with oxidizing agents and strong acids such as sulfuric and nitric acids.

Ingestion of sucrose does not cause harm to a healthy individual, as it is a completely digestible food. However, sucrose intake is restricted in some diabetic patients due to its glycemic index of 65, which prompts the pancreas to secrete insulin to lower blood glucose levels (Veit et al., 2022).

	SUGAR	
	Other names: Sucrose, saccharose, white sugar, table sugar, and α-D-glucopyranosyl-β-D-fructofuranoside	
	Origin: Natural	Formula: $C_{12}H_{22}O_{11}$
Type: Carbohydrate, sugars, disaccharide	Discoverer: Unknown	MW: 342.3 g/mol
Appearance: White or yellow crystals	Inventor: NA	mp: decomposes at 186°C
Uses: Food and food additive	Toxicity: No	bp: NA

The Structure and Characterization of Sugar

Sucrose is a simple sugar; therefore, it is necessary to describe the family of simple sugars, or carbohydrates, composed of monosaccharides and disaccharides. Monosaccharides with six carbon (C) atoms are called hexoses ($C_6H_{12}O_6$) and contain five hydroxyl (OH) groups. The spatial arrangement of hexoses results in different isomers; in theory, there are 16 aldohexose isomers and 8 2-ketohexose isomers (Capon, 1969).

Aldohexoses have four chiral centers. By applying the mathematical expression 2^n, the maximum number of isomers can be calculated, where n represents the number of chiral centers. In this case, substituting n = 4 yields $2^4 = 16$. The notation D- (dextro) and L- (levo) is used to differentiate enantiomers based on their spatial arrangement. These isomers are grouped into 8 pairs of D- and L- enantiomers, named allose, altrose, glucose, mannose, gulose, idose, galactose, and talose, respectively. Only the D-enantiomers are found in nature, with D-glucose being the most abundant.

Since there are three chiral centers in 2-ketohexoses, and according to the mathematical expression 2^n, the number of isomers is $2^3 = 8$. The 8 isomers are divided into 4 pairs of D- and L- enantiomers: psicose, fructose, sorbose, and tagatose. The most common among them is D-fructose, found in honey and fruits in general (Akram & Hamid, 2013).

It is worth mentioning that pentoses are also sugars; they contain five carbon (C) atoms in their structure ($C_5H_{10}O_5$). With three chiral centers, they form a total of eight isomers, divided into four pairs of D- and L-enantiomers: ribose, arabinose, xylose, and lyxose (Ruchala & Sibirny, 2021).

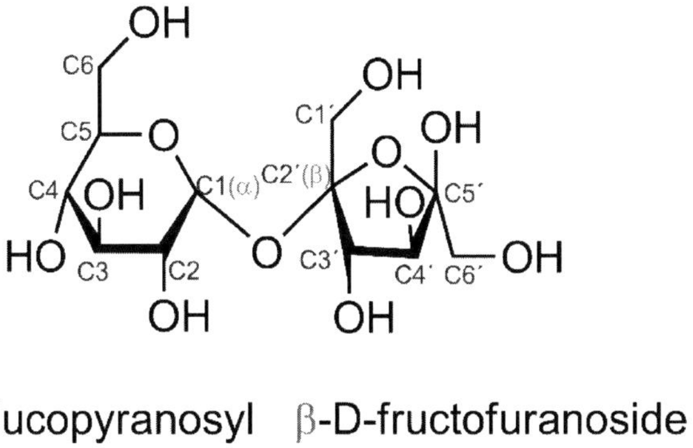

α-D-glucopyranosyl β-D-fructofuranoside

Figure 1. Sucrose structure and scientific name with carbon atoms labeled.

Hexoses can be represented in their closed hemiacetal ring form, then, the first carbon is numbered as C1 and is assigned to the center linked to the two O atoms. The remaining carbon atoms are numbered consecutively from C2 to C6. The stereoisomers are designated as α (alpha) and β (beta) according to the configuration of the -OH group bonded in the anomeric center C1; the OH group in α anomer is below the plane of the ring, while in β anomer OH is above the plane (Pérez, 1995). Glucose undergoes anomeric mutarotation in solution due to ring opening and closing. In sucrose, the subunits are linked via a glycosidic bond between C1 on the glucosyl subunit and C2 on the fructosyl subunit, and only the α-D-glucopyranosyl and β-D-fructofuranosyl configuration is observed, without mutarotation (Figure 1).

The structure of sucrose was confirmed in the 20th century by X-ray diffraction when the structure of sucrose sodium bromide dihydrate was published in 1947 (Beevers & Cochran, 1947), and later, when the structure of pure sucrose was solved in 1952 (Beevers et al., 1952). Years later, Brown and Levy, in 1963, presented a more precise refinement of sucrose using neutron diffraction, in which the cell volume was determined to be 715 $Å^3$ and the space group $P2_1$, finding that all O atoms are involved in H-bonding, with both intermolecular and intramolecular interactions (Pérez, 1995). The characterization of sucrose by ^{1}H-NMR in D_2O (Chemical Book, 2024) shows the number of signals in the expected region for OH and aliphatic H, respectively; the chemical shifts observed are 5.42, 4.22, 4.05, 3.89, 3.86, 3.63, 3.82, 3.76, 3.68, 3.56, and 3.47 ppm. The infrared spectrum shows a broad band at 3338 cm^{-1}, corresponding to the OH groups, followed by several bands around 2943 cm^{-1}, corresponding to the C-H stretching of aliphatic rings.

Obtaining Sugar

Table sugar is a compound biosynthesized by plants as part of their metabolism. Plants derive their nutrients from carbon sources, such as carbon dioxide (CO_2) and other substances from substrates. Additionally, plants absorb water and trace elements from soil salts and minerals. Through these processes, plants biosynthesize monosaccharides (e.g., glucose), disaccharides (e.g., sucrose), and polysaccharides (e.g., starch) (Stirbet et al., 2020). Chloroplasts undergo photosynthesis in two stages (Figure 2.1) (Gan et al., 2019). The first stage of photosynthesis is the light reaction (Figure 2.2), in which adenosine diphosphate (ADP) and nicotinamide adenine dinucleotide phosphate ($NADP^+$) utilize light and water to produce ATP and reduced nicotinamide adenine dinucleotide phosphate (NADPH) via photosystems II and I in the electron transport chain, occurring in the granum lamellae and thylakoids. Oxygen is released into the atmosphere during the first stage. In the second stage of photosynthesis, referred to as the dark reaction (Figure 2.3), ATP and NADPH fix CO_2 from the surroundings to produce carbohydrates. The Calvin cycle explains the mechanisms underlying these reactions, in which the 3-carbon compound glyceraldehyde-3-phosphate produces glucose, fructose, sucrose, starch, and other carbohydrates (Salt Lake Community College, 2024).

This biosynthesis process is not exclusive to plants; some microorganisms, such as cyanobacteria and proteobacteria, also fix CO_2, transforming it into simple sugars (Lunn, 2002).

Following sugar biosynthesis, the human hand comes into action to extract and refine the compound. The first step is the cultivation of crops; in the case of sugarcane, it is harvested and transported to the sugar mill for further processing (Panigrahi et al., 2021). Once in the factory, the cane stalks are crushed and chopped. Water is added to maximize the extraction of saccharose; then, the solution is filtered to separate solid residues, and the concentrated sugar juice is heated to eliminate microorganisms and reduce excess water. The mixture is sent to a crystallization chamber, where small amounts of sugar crystals are added to promote crystallization; saccharose and molasses mix at this stage. Saccharose and molasses are centrifuged at approximately 1200 rpm to separate them; molasses remains in the liquid phase and saccharose crystals in the centrifuge. Both molasses and saccharose are used in various consumer goods.

$$\text{(1)}\quad 6CO_2 + H_2O \xrightarrow{\text{Photosynthesis}} \text{Carbohydrates} + O_2$$

$$\text{(2)}\quad ADP + NADP^+ + H_2O \xrightarrow[\text{GL/ETC/Thy}]{\text{"Light reaction"}} ATP + NADPH + O_2$$

$$\text{(3)}\quad 6CO_2 + ATP + NADPH \xrightarrow[\text{Calvin cycle}]{\text{"Dark reaction"}} \text{Carbohydrates} + ADP + NADP^+$$

GL = Grana lamellae
ETC = Electron transport chain
Thy = Thylakoids
Carbohydrates = Glucose, sucrose, starch, etc.

Figure 2. Reaction schemes for CO_2 fixation during photosynthesis (1), the light reaction (2), and the dark reaction (3) of photosynthesis.

The synthesis of sucrose has been successfully achieved in the laboratory; however, alternative routes cannot compete with natural table sugar. Nonetheless, synthetic alternatives demonstrate the feasibility of obtaining novel disaccharides that are not found in nature (Jarosz et al., 2020). The first study on synthetic sucrose was published in 1953 by Canadian researchers Raymond Lemieux (1920–2000) and George Huber, who reported a methodology with a yield of 5.5% (Lemieux & Huber, 1953). Subsequent investigations on the total synthesis of sucrose overcame the difficulties of directing reactions toward isolating a single isomer (Borman, 2000). Coinciding, in 2000, the year of Dr. Lemieux's passing, Stefan Oscarson and Fernando Sehgelmeble from Sweden presented the stereospecific synthesis of several disaccharides, including sucrose with an 80% yield (Oscarson & Sehgelmeble, 2000).

The Sugar Industry

Sugar is an essential ingredient in many food products. It is crucial for the confectionery and baking industries, as well as for the soft drink and non-carbonated beverage industries (e.g., juices, flavored milk), where syrups and concentrates used during processing contain high amounts of simple sugars. Furthermore, sugar, along with its derivatives glucose and fructose, serves as a raw material in other industrial sectors, such as fermented products. For example, sugars and brewer's yeast (*Saccharomyces cerevisiae*) are the main ingredients in the production of alcoholic beverages (Sicard & Legras, 2011). Another important disaccharide in the food industry is lactose, a compound found in milk and utilized to produce dairy products such as yogurt and cheese through fermentation by lactic acid bacteria (Dominici et al., 2022). Because sucrose is a safe food, it is also utilized in the pharmaceutical industry as an excipient or non-active ingredient in pills, tablets, and medical syrups (Strickley & Lambert, 2021).

Several alternatives to sucrose exist, including high-fructose corn syrup, a sweetener composed of approximately equal parts of the monomers D-glucose and D-fructose (Ashurst & Hargitt, 2009). Other natural sweeteners include xylitol, agave nectar, stevia, honey, yacon syrup, monk fruit, and coconut sugar (Gardner, 2017).

There are also artificial sweetener alternatives such as sucralose, saccharin, aspartame, xylitol, acesulfame-K, neotame, and advantame (Ghusn et al., 2023).

Artificial sweeteners have pros and cons. The main benefit of artificial sweeteners is their low caloric content, as sucrose provides many calories—approximately 387 cal per 100 grams (Nutritionix, 2024). Artificial substitutes address this issue by providing greater sweetening power while adding very few calories to the diet (Dominici et al., 2022). Unfortunately, this benefit is not without drawbacks, as explained below.

The list of disadvantages of artificial sucrose substitutes is quite long. The first drawback is the cost; these products are usually more expensive; therefore, not everyone has access to them. Additionally, there are negative health-related aspects. Decades ago, artificial sweeteners were considered suitable alternatives for people diagnosed with obesity or type 2 diabetes, since their caloric intake was low or negligible. However, recent studies have found that chronic consumption of these sweeteners involves certain complications (Kossiva et al., 2024).

One significant health concern is that regular intake of artificial sweeteners may cause an imbalance in the gut microbiome (Iizuka, 2022). Another issue is that artificial sweeteners can induce glucose intolerance (Suez et al., 2014). Finally, the most serious health risks associated with artificial sweeteners involve cancer. For example, sodium cyclamate has been banned by the FDA since the 1970s because it increases the risk of cancer and birth defects in neonates. In 2023, aspartame was classified by the International Agency for Research on Cancer (IARC) as a substance possibly carcinogenic to humans (Group 2B) (American Cancer Society, 2024). Furthermore, independent studies have also associated aspartame with certain types of cancer when consumed in high quantities (Palomar-Cros et al., 2023).

Therefore, artificial sweeteners should be consumed in moderation and avoided in daily diets. Natural sugar substitutes are more viable, as they are competitively priced and offer a similar sweetening power to sugar.

Sugar Production Around the World

The world's top 10 sugarcane-producing countries in 2022 (Our World in Data, 2024) were: Brazil (724 million metric tons), India (439 million metric tons), China (103 million metric tons), Thailand (92 million metric tons), Pakistan (88 million metric tons), Mexico (55 million metric tons), Colombia (35 million metric tons), Indonesia (32 million metric tons), the USA (31 million metric tons), and Australia (29 million metric tons).

Global sugar consumption has been increasing consistently in recent years, rising from 155.8 million metric tons in 2010 to approximately 178 million metric tons in 2024 (Statista, 2024). It is projected to continue rising in the coming years, with India as the largest sugar consumer.

What If Sugar Is No Longer Produced?

The complete halt in table sugar production would require strong justification due to its economic repercussions. For example, if sugar exports were banned in a sugar-producing country, the harvest volume in that hypothetical region would decrease, as the surplus could not be sold. This situation would negatively impact on the local economy, as numerous jobs across the supply chain—such as farmers, laborers, and transporters—would be lost. Nonetheless, if a country under export restriction conditions continues to produce a high volume of sugar, the local price would decrease, at least in the short term due to the law of supply and demand. Currently, in some countries, there are restrictions or prohibitions on the export of sugarcane and table sugar, but for different reasons and with different consequences. Two cases are mentioned below.

The first case occurred recently when, in 2023, the Indian government restricted sugarcane and table sugar exports to control prices and ensure domestic availability, in response to the growing global demand for bioethanol fuels (Rajput & Venkataraman, 2024).

The second case began a long time ago, in 1962, when the US government announced an embargo on Cuba due to economic and political differences (Gordon, 2016). Notably, the embargo affected not only sugar but also the trade of other products. Nonetheless, it is worth remembering that Cuba was once a leading exporter of table sugar; however, the embargo, combined with other economic and political factors, contributed to the decline in sugarcane production. Output fell from 82 million metric tons in 1990 to 36 million metric tons in 2000, and further to only 11 million metric tons in 2020 (Our World in Data, 2024).

As it is seen while India's restrictions aimed to stabilize prices, Cuba's embargo had longer-lasting economic and political consequences.

But what if sugar production is not only restricted but fully banned worldwide? In such a scenario, the impact would be profound and far-reaching. A cascade effect would begin by affecting all local and transnational companies dependent on sugar raw materials, such as soft drink and confectionery manufacturers. Given that these companies bear a high tax burden, governments would experience a loss of tax revenue, thereby affecting their economies. In addition, the black market would trade sugar at a higher price, resulting in effects comparable to those of illegal drug trafficking. Given these assumptions, estimating the full extent of collateral effects is almost impossible. Fortunately, it is highly unlikely that table sugar and its derivative products would be removed from supermarkets.

An Opinion About Sugar

Table sugar is a dangerous ally because this valued food carries a negative stigma that cannot be ignored. Demographic growth leads to an increase in crop production to satisfy consumption needs. However, although the food produced worldwide currently exceeds demand, it is not distributed equitably, creating an imbalance in caloric intake per individual. While one part of the population consumes excessive calories, another part suffers from insufficient caloric intake (Bratanova et al., 2016). Nevertheless, excessive sugar intake is independent of factors such as socioeconomic status, race, age, or gender (Thompson et al., 2009), which is a worrying situation.

Table sugar is a high-calorie food, providing approximately 387 calories per 100 grams (Nutritionix, 2024), so sugar can contribute positively to the diet when consumed in appropriate portions. However, when consumption is excessive, nutritional problems may arise. The digestive system can tolerate excesses or deficiencies of certain foods, but long-term excessive sugar intake can lead to health problems such as insulin resistance, diabetes, and obesity (Chandrasekaran & Weiskirchen, 2024; Macdonald, 2016), leading to further complications that may compromise health temporarily or permanently.

Despite the issues associated with excessive sugar intake, it is commonly found in junk foods to enhance their appeal. The problem is that table sugar consumption affects the central nervous system in a manner comparable to certain drugs by triggering dopamine release (Rada et al., 2005; Yudkin, 1972). The food industry seizes the need for sugar to increase sales. Moreover, sugar is added to tasteless (processed) foods to make them a calorie trap.

Sugar is not the sole cause of obesity and diabetes in the population; however, uncontrolled intake may contribute to the development of these conditions, which affect health. Unfortunately, it is easy to exceed recommended portions because many foods contain added sugars, and a lack of information about portion sizes contributes to overconsumption. Certain products, such as large soft drink bottles (with a capacity of 32 oz or 1 L and larger) and large ice cream packs, clearly exceed the daily recommended portion, yet most people remain unaware of this. Additionally, another dangerous factor is addiction, as some studies indicate that sugar is a highly addictive compound (Witek et al., 2022). This may explain why eliminating sugar from the diet is so difficult, despite warnings about its health risks.

The conservative detractors point out that consuming refined sugar has more disadvantages than advantages and suggest that it is better to obtain simple carbohydrates from natural sources such as fruits (Nebraska Medicine, 2024). Other researchers take a stricter stance on sugar consumption, as they also advocate reducing the percentage of total carbohydrates in the diet to lower the risk of heart problems, obesity, and diabetes (Oh et al., 2023).

Research on low-carb diets suggests that healthy nutrition can be achieved by reducing carbohydrate intake and replacing it with other food groups such as proteins or fats. They argue that only a small percentage of people lead a nomadic lifestyle compared to those in urban and rural areas, where sedentary living is predominant (Alves et al., 2016). For example, an analysis of young women with a sedentary lifestyle who followed a low-carbohydrate diet combined with exercise showed positive effects, including improved cardiometabolic health and reduced anxiety levels (Hu et al., 2022). While researchers who advocate ketogenic diets take a more extreme approach, proposing a total ban on carbohydrates, including fruits (O'Neill & Raggi, 2020).

Types of diets are still a topic of discussion among experts; for now, it is up to each person to select the nutrients they consume, despite the lack of robust studies. Due to the differences between conflicting dietary approaches, it is advisable to tailor caloric intake to individual needs. A well-balanced diet should consider factors such as age, gender, lifestyle, physical activity, and health status. In this way, it becomes easier to assess whether consuming foods such as table sugar is appropriate and, if so, to determine the correct amount for maintaining a healthy lifestyle.

Humans are not the only ones affected by the intensified planting in monocultures; the environment also suffers negative consequences. The cultivation of raw materials for table sugar requires high volumes of water, fertilizers, and pesticides (Rad et al., 2022). Given that the demand for table sugar increases each year, a larger allocation of land is designated for monoculture. Preparation of farmlands often causes both legal and illegal deforestation, which negatively impacts wildlife (Holland, 2004), soil, and even subsoil (Suarez & Gwozdz, 2023). Therefore, this is another reason to be mindful of the type and quantity of food we consume.

Final Comments on Sugar

The intake of sugar provides immediate energy required for performing high-intensity physical activities and is essential for the proper functioning of the body when consumed in moderation. Its availability and caloric value provide sufficient justification for recognizing table sugar as a milestone food, as having a daily meal—even if it includes some junk food—is preferable to not eating at all.

Some Curiosities About Sugars

- A cave painting depicting a honey collector climbing a tree, known as the Man of Bicorp, is located in La Cueva de la Araña (The Spider Cave) in Valencia, Spain, and dates back to approximately 15,000 BC (Hajar, 2008).

- In ancient Egypt, Greece, and Rome, honey was used to treat wounds and accelerate healing (Kuropatnicki et al., 2018). Modern medicine has resumed using pure honey in treatments because, beyond its antiseptic properties, it promotes epithelial regeneration (Ranzato et al., 2012).
- The invention of the sugar cube in 1843 is credited to the Swiss couple Jacob Christoph Rad and his wife, Juliana Rad, who had the idea of cutting sugar into small cubes as a practical alternative to the long sugar loaves that were sold at the time as the standard form (Potthast, 2025).
- It is well known that flying insects like bees and wasps collect nectar to create honeycombs and nests, respectively, storing food for future use or to feed their larvae. However, certain genera of honey ants, such as *Myrmecocystus* (van Elst et al., 2021) and *Camponotus* (Islam et al., 2022), also collect nectar but store it within their bodies rather than in external structures.
- Bee honey is composed of varying carbohydrates. The primary components found in blossom honey are as follows: fructose (38%), glucose (31%), water (17%), other oligosaccharides (4%), other disaccharides (5%), erlose (0.8%), and sucrose (0.7%) (Puścion-Jakubik et al., 2020). Unfortunately, pure honey is adulterated by adding sucrose.
- Table sugar resists microbial growth and does not require refrigeration. It has a long shelf life and can typically be consumed up to two years after production (Utah State University, 2015).
- Sucrose exhibits triboluminescence, meaning it emits light when subjected to mechanical stress (Zink et al., 1976).
- The sweet taste of sugar is detected at the tip of the tongue, while salty, sour, bitter, and umami flavors are perceived in other areas (Spence, 2022).
- The total sugars (sucrose, fructose, or glucose) in a can of Coca-Cola (355 mL or 12 oz) are approximately 39 g (The Coca-Cola company, 2025). This amount varies by country, but it exceeds the 32 g of total added sugars recommended in a 2,000-calorie diet (Johnson et al., 2009).
- Sugar consumption increases levels of the neurotransmitter dopamine, which plays a role in producing a feeling of well-being (Jacques et al., 2019). This stimulation can lead to an increase in sugar

intake, potentially contributing to health issues such as obesity and diabetes. Similarly, dopamine is released when consuming certain addictive substances, including ethanol (Söderpalm & Ericson, 2024) or cocaine (Yablonski-Alter et al., 2008).

- Confirmation of a fasting plasma glucose test result above 126 mg/dL is consistent with a diagnosis of diabetes (Kaur et al., 2020).
- *The mellified man* is a legendary and aberrant medicinal practice mentioned in the book *Chinese Materia Medica*, compiled by the naturalist Li Shih-Chen in 1597. This process allegedly begins when a person close to death is fed only honey. Upon their passing, the corpse is immersed in a sarcophagus filled with honey, which is then sealed and left to macerate for at least a century. When the sarcophagus is eventually opened, the substance obtained is believed to have acquired medicinal properties (Roach, 2003). However, there is no evidence that this type of cannibalistic concoction was ever made. It is likely a misunderstanding, as in ancient times, food offerings—including containers of honey—were commonly placed in tombs and burial sites (Chichinadze et al., 2019; Rosiak et al., 2024).
- The words honey and honeycomb appear 61 times in the 21st Century King James Version. The first time in Genesis 43:11, "And their father Israel said unto them, If it must be so now, do this: Take of the best fruits in the land in your vessels, and carry down the man a present, a little balm and a little honey, spices and myrrh, nuts and almonds" (*Holy Bible. The 21st Century King James Version*, 1994).

References

Akram, M., & Hamid, A. (2013). Mini review on fructose metabolism. *Obesity Research & Clinical Practice*, *7*(2), e89–e94. https://doi.org/10.1016/j.orcp.2012.11.002.

Alves, B., Silva, T., & Spritzer, P. (2016). Sedentary lifestyle and high-carbohydrate intake are associated with low-grade chronic inflammation in post-menopause: A cross-sectional study. *Revista Brasileira de Ginecologia e Obstetrícia*, *38*(07), 317–324. https://doi.org/10.1055/s-0036-1584582.

American Cancer Society. (2024, November 15). *Aspartame and Cancer Risk*. Https://Www.Cancer.Org/Cancer/Risk-Prevention/Chemicals/Aspartame.Html.

Ashurst, P. R., & Hargitt, R. (2009). Ingredients in soft drinks and fruit juices. In P. R. Ashurst, R. Hargitt, & F. Palmer (Eds.), *Soft Drink and Fruit Juice Problems Solved* (pp. 20–59). Elsevier. https://doi.org/10.1533/9781845697068.20.

Beevers, C. A., & Cochran, W. (1947). The crystal structure of sucrose sodium bromide dihydrate. *Proceedings of the Royal Society of London. Series A. Mathematical and Physical Sciences*, *190*(1021), 257–272. https://doi.org/10.1098/rspa.1947.0075.

Beevers, C. A., McDonald, T. R. R., Robertson, J. H., & Stern, F. (1952). The crystal structure of sucrose. *Acta Crystallographica*, *5*(5), 689–690. https://doi.org/10.1107/S0365110X52001908.

Borman, S. (2000, September 25). Sucrose synthesis sets a record. *Chemical & Engineering News Archive*, 52. https://doi.org/10.1021/cen-v078n039.p052.

Boutros, A. R. (1962). Carbohydrate metabolism-a review. *Canadian Anaesthetists' Society Journal*, *9*(4), 353–370. https://doi.org/10.1007/BF03021273.

Bratanova, B., Loughnan, S., Klein, O., Claassen, A., & Wood, R. (2016). Poverty, inequality, and increased consumption of high calorie food: Experimental evidence for a causal link. *Appetite*, *100*, 162–171. https://doi.org/10.1016/j.appet.2016.01.028.

Canadian Sugar Institute. (2024, November 15). *Sources of sugar*. Https://Sugar.ca/Sugar-Basics/Sources-of-Sugar.

Capon, B. (1969). Mechanism in carbohydrate chemistry. *Chemical Reviews*, *69*(4), 407–498. https://doi.org/10.1021/cr60260a001.

Chandrasekaran, P., & Weiskirchen, R. (2024). The role of obesity in type 2 diabetes mellitus—an overview. *International Journal of Molecular Sciences*, *25*(3), 1882. https://doi.org/10.3390/ijms25031882.

Chemical Book. (2024, November 15). *Sucrose(57-50-1) 1H NMR*. Https://Www.Chemicalbook.Com/SpectrumEN_57-50-1_1HNMR.Htm.

Chichinadze, M., Kvavadze, E., Martkoplishvili, I., & Kacharava, D. (2019). Palynological evidence for the use of honey in funerary rites during the Classical Period at the Vani. *Quaternary International*, *507*, 24–33. https://doi.org/10.1016/j.quaint.2019.01.011.

Combes, D., & Monsan, P. (1983). Sucrose hydrolysis by invertase. Characterization of products and substrate inhibition. *Carbohydrate Research*, *117*, 215–228. https://doi.org/10.1016/0008-6215(83)88088-4.

Dinesh Babu, K. S., Janakiraman, V., Palaniswamy, H., Kasirajan, L., Gomathi, R., & Ramkumar, T. R. (2022). A short review on sugarcane: its domestication, molecular manipulations and future perspectives. *Genetic Resources and Crop Evolution*, *69*(8), 2623–2643. https://doi.org/10.1007/s10722-022-01430-6.

Dominici, S., Marescotti, F., Sanmartin, C., Macaluso, M., Taglieri, I., Venturi, F., Zinnai, A., & Facioni, M. S. (2022). Lactose: Characteristics, food and drug-related applications, and its possible substitutions in meeting the needs of people with lactose intolerance. *Foods*, *11*(10), 1486. https://doi.org/10.3390/foods11101486.

Gan, P., Liu, F., Li, R., Wang, S., & Luo, J. (2019). Chloroplasts— beyond energy capture and carbon fixation: Tuning of photosynthesis in response to chilling stress. *International Journal of Molecular Sciences*, *20*(20), 5046. https://doi.org/10.3390/ijms20205046.

Gardner, E. (2017). The unbearable sweetness of sugar (and sugar alternatives). *BDJ Team*, *4*(9), 17156. https://doi.org/10.1038/bdjteam.2017.156.

Ghusn, W., Naik, R., & Yibrin, M. (2023). The impact of artificial sweeteners on human health and cancer association: A comprehensive clinical review. *Cureus*. https://doi.org/10.7759/cureus.51299.

Gordon, J. (2016). Economic sanctions as 'negative development': The case of Cuba. *Journal of International Development*, *28*(4), 473–484. https://doi.org/10.1002/jid.3061.

Hajar, R. (2008). Honey and medicine. In H. Selin (Ed.), *Encyclopaedia of the History of Science, Technology, and Medicine in Non-Western Cultures* (2nd ed., pp. 1074–1079). Springer Netherlands. https://doi.org/10.1007/978-1-4020-4425-0_9705.

Harveson, R. M. (2024, November 15). *History of sugarbeets*. Https://Cropwatch.Unl.Edu/History-Sugarbeets.

Holland, R. (2004, November 22). *Tackling the sweets that cause ecological decay*. Https://Wwf.Panda.Org/Wwf_news/?16670/Tackling-the-Sweets-That-Cause-Ecological-Decay.

Holy Bible. The 21st Century King James Version. (1994). Deuel Enterprises.

Hu, M., Shi, Q., Sun, S., Hong, H. I., Zhang, H., Qi, F., Zou, L., & Nie, J. (2022). Effect of a low-carbohydrate diet with or without exercise on anxiety and eating behavior and associated changes in cardiometabolic health in overweight young women. *Frontiers in Nutrition*, *9*. https://doi.org/10.3389/fnut.2022.894916.

Iizuka, K. (2022). Is the use of artificial sweeteners beneficial for patients with diabetes Mellitus? The Advantages and Disadvantages of Artificial Sweeteners. *Nutrients*, *14*(21), 4446. https://doi.org/10.3390/nu14214446.

Islam, M. K., Lawag, I. L., Sostaric, T., Ulrich, E., Ulrich, D., Dewar, T., Lim, L. Y., & Locher, C. (2022). Australian honeypot ant (Camponotus inflatus) honey—A comprehensive analysis of the physiochemical characteristics, bioactivity, and HPTLC profile of a traditional indigenous Australian food. *Molecules*, *27*(7), 2154. https://doi.org/10.3390/molecules27072154.

Jacques, A., Chaaya, N., Beecher, K., Ali, S. A., Belmer, A., & Bartlett, S. (2019). The impact of sugar consumption on stress driven, emotional and addictive behaviors. *Neuroscience & Biobehavioral Reviews*, *103*, 178–199. https://doi.org/10.1016/j.neubiorev.2019.05.021.

Jarosz, S., Sokołowska, P., & Szyszka, Ł. (2020). Synthesis of fine chemicals with high added value from sucrose: Towards sucrose-based macrocycles. *Tetrahedron Letters*, *61*(22), 151888. https://doi.org/10.1016/j.tetlet.2020.151888.

Johnson, R. K., Appel, L. J., Brands, M., Howard, B. V., Lefevre, M., Lustig, R. H., Sacks, F., Steffen, L. M., & Wylie-Rosett, J. (2009). Dietary sugars intake and cardiovascular health. *Circulation*, *120*(11), 1011–1020. https://doi.org/10.1161/CIRCULATIONAHA.109.192627.

Kaur, G., Lakshmi, P. V. M., Rastogi, A., Bhansali, A., Jain, S., Teerawattananon, Y., Bano, H., & Prinja, S. (2020). Diagnostic accuracy of tests for type 2 diabetes and prediabetes: A systematic review and meta-analysis. *PLOS ONE*, *15*(11), e0242415. https://doi.org/10.1371/journal.pone.0242415.

Kossiva, L., Kakleas, K., Christodouli, F., Soldatou, A., Karanasios, S., & Karavanaki, K. (2024). Chronic use of artificial sweeteners: Pros and cons. *Nutrients*, *16*(18), 3162. https://doi.org/10.3390/nu16183162.

Kuropatnicki, A. K., Kłósek, M., & Kucharzewski, M. (2018). Honey as medicine: historical perspectives. *Journal of Apicultural Research*, *57*(1), 113–118. https://doi.org/10.1080/00218839.2017.1411182.

Lemieux, R. U., & Huber, G. (1953). A chemical synthesis of sucrose. *Journal of the American Chemical Society*, *75*(16), 4118–4118. https://doi.org/10.1021/ja01112a545.

Lunn, J. E. (2002). Evolution of sucrose synthesis. *Plant Physiology*, *128*(4), 1490–1500. https://doi.org/10.1104/pp.010898.

Macdonald, I. A. (2016). A review of recent evidence relating to sugars, insulin resistance and diabetes. *European Journal of Nutrition*, *55*(S2), 17–23. https://doi.org/10.1007/s00394-016-1340-8.

Nebraska Medicine, U. H. C. (2024, November 15). *Sugar in fruits: What matters for health*. Https://Health.Unl.Edu/Sugar-Fruits-What-Matters-Health.

Nutritionix. (2024, November 15). *100 G Sugar*. Https://Www.Nutritionix.Com/Food/Sugar/100-g.

Oh, R., Gilani, B., & Uppaluri, K. R. (2023, August 17). *Low-Carbohydrate Diet*. StatPearls. https://www.ncbi.nlm.nih.gov/books/NBK537084/.

O'Neill, B., & Raggi, P. (2020). The ketogenic diet: Pros and cons. *Atherosclerosis*, *292*, 119–126. https://doi.org/10.1016/j.atherosclerosis.2019.11.021.

Oscarson, S., & Sehgelmeble, F. W. (2000). A novel β-directing fructofuranosyl donor concept. Stereospecific synthesis of sucrose. *Journal of the American Chemical Society*, *122*(37), 8869–8872. https://doi.org/10.1021/ja001439u.

Our World in Data. (2024, November 15). *Sugar cane production*. Https://Ourworldindata.Org/Grapher/Sugar-Cane-Production.

Palomar-Cros, A., Straif, K., Romaguera, D., Aragonés, N., Castaño-Vinyals, G., Martin, V., Moreno, V., Gómez-Acebo, I., Guevara, M., Aizpurua, A., Molina-Barceló, A., Jiménez-Moleón, J., Tardón, A., Contreras-Llanes, M., Marcos-Gragera, R., Huerta, J. M., Pérez-Gómez, B., Espinosa, A., Hernández-Segura, N., ... Lassale, C. (2023). Consumption of aspartame and other artificial sweeteners and risk of cancer in the Spanish multicase-control study (MCC-Spain). *International Journal of Cancer*, *153*(5), 979–993. https://doi.org/10.1002/ijc.34577.

Panigrahi, C., Shaikh, A. E. Y., Bag, B. B., Mishra, H. N., & De, S. (2021). A technological review on processing of sugarcane juice: Spoilage, preservation, storage, and packaging aspects. *Journal of Food Process Engineering*, *44*(6). https://doi.org/10.1111/jfpe.13706.

Pérez, S. (1995). The structure of sucrose in the crystal and in solution. In M. M. P. Reiser (Ed.), *Sucrose* (pp. 11–32). Springer US. https://doi.org/10.1007/978-1-4615-2676-6_2.

Potthast, J. B. (2025, January 26). *Cube Sugar An Invention that Sweetened Everyday Life*. Https://Www.Dpma.de/English/Our_office/Publications/Milestones/Inventionsthatmadehistory/Sugarcubes/Index.Html.

Puścion-Jakubik, A., Borawska, M. H., & Socha, K. (2020). Modern methods for assessing the quality of bee honey and botanical origin identification. *Foods*, *9*(8), 1028. https://doi.org/10.3390/foods9081028.

Rad, S. M., Ray, A. K., & Barghi, S. (2022). Water pollution and agriculture pesticide. *Clean Technologies*, *4*(4), 1088–1102. https://doi.org/10.3390/cleantechnol4040066.

Rada, P., Avena, N. M., & Hoebel, B. G. (2005). Daily bingeing on sugar repeatedly releases dopamine in the accumbens shell. *Neuroscience*, *134*(3), 737–744. https://doi.org/10.1016/j.neuroscience.2005.04.043.

Rajput, R., & Venkataraman, S. V. (2024). Revenue-sharing contract with government subsidy: A case of the Indian sugar supply chain. *Computers & Industrial Engineering*, *191*, 110159. https://doi.org/10.1016/j.cie.2024.110159.

Ranzato, E., Martinotti, S., & Burlando, B. (2012). Epithelial mesenchymal transition traits in honey-driven keratinocyte wound healing: Comparison among different honeys. *Wound Repair and Regeneration*, *20*(5), 778–785. https://doi.org/10.1111/j.1524-475X.2012.00825.x.

Roach, M. (2003). *Stiff: The Curious Lives of Human Cadavers* (1st ed.). W. W. Norton & Company.

Rosiak, A., Józefowska, A., Sekulska-Nalewajko, J., Gocławski, J., & Kałużna-Czaplińska, J. (2024). Funerary vs. domestic vessels from the Hallstatt period. A study on ceramic vases from the Milejowice settlement and the Domasław cemetery. *Scientific Reports*, *14*(1), 19942. https://doi.org/10.1038/s41598-024-70219-7.

Ruchala, J., & Sibirny, A. A. (2021). Pentose metabolism and conversion to biofuels and high-value chemicals in yeasts. *FEMS Microbiology Reviews*, *45*(4). https://doi.org/ 10.1093/femsre/fuaa069.

Salt Lake Community College. (2024, December 20). *The Calvin Cycle*. Https://Slcc.Pressbooks.Pub/Collegebiology1/Chapter/the-Calvin-Cycle/.

Sicard, D., & Legras, J.-L. (2011). Bread, beer and wine: Yeast domestication in the Saccharomyces sensu stricto complex. *Comptes Rendus. Biologies*, *334*(3), 229–236. https://doi.org/10.1016/j.crvi.2010.12.016.

Söderpalm, B., & Ericson, M. (2024). Alcohol and the dopamine system. In A. de Bejczy & B. Söderpalm (Eds.), *The neurobiology of alcohol use disorder: Neuronal mechanisms, current treatments and novel developments* (Vol. 175, pp. 21–73). Elsevier. https://doi.org/10.1016/bs.irn.2024.02.003.

Spence, C. (2022). The tongue map and the spatial modulation of taste perception. *Current Research in Food Science*, *5*, 598–610. https://doi.org/10.1016/j.crfs.2022.02.004.

Statista. (2024, November 15). *Sugar Industry - statistics & facts*. Https://Www. Statista.Com/Topics/1224/Sugar/#topicOverview.

Stirbet, A., Lazár, D., Guo, Y., & Govindjee, G. (2020). Photosynthesis: basics, history and modelling. *Annals of Botany*, *126*(4), 511–537. https://doi.org/10.1093/aob/mcz171.

Strickley, R. G., & Lambert, W. J. (2021). A review of formulations of commercially available antibodies. *Journal of Pharmaceutical Sciences*, *110*(7), 2590-2608.e56. https://doi.org/10.1016/j.xphs.2021.03.017.

Suarez, A., & Gwozdz, W. (2023). On the relation between monocultures and ecosystem services in the Global South: A review. *Biological Conservation*, *278*, 109870. https://doi.org/10.1016/j.biocon.2022.109870.

Suez, J., Korem, T., Zeevi, D., Zilberman-Schapira, G., Thaiss, C. A., Maza, O., Israeli, D., Zmora, N., Gilad, S., Weinberger, A., Kuperman, Y., Harmelin, A., Kolodkin-Gal, I., Shapiro, H., Halpern, Z., Segal, E., & Elinav, E. (2014). Artificial sweeteners induce glucose intolerance by altering the gut microbiota. *Nature*, *514*(7521), 181–186. https://doi.org/10.1038/nature13793.

The Coca-Cola company. (2025, January 26). *How much sugar is in Coca-Cola?* Https://Www.Coca-Colacompany.Com/about-Us/Faq/How-Much-Sugar-Is-in-Coca-Cola.

Thompson, F. E., McNeel, T. S., Dowling, E. C., Midthune, D., Morrissette, M., & Zeruto, C. A. (2009). Interrelationships of added sugars intake, socioeconomic status, and race/ethnicity in adults in the United States: National health interview survey, 2005. *Journal of the American Dietetic Association, 109*(8), 1376–1383. https://doi.org/10.1016/j.jada.2009.05.002.

Utah State University. (2015, January 25). *Storing Sugars*. Https://Extension.Usu.Edu/Preserve-the-Harvest/Research/Storing-Sugars.

van Elst, T., Eriksson, T. H., Gadau, J., Johnson, R. A., Rabeling, C., Taylor, J. E., & Borowiec, M. L. (2021). Comprehensive phylogeny of Myrmecocystus honey ants highlights cryptic diversity and infers evolution during aridification of the American Southwest. *Molecular Phylogenetics and Evolution, 155*, 107036. https://doi.org/10.1016/j.ympev.2020.107036.

Veit, M., van Asten, R., Olie, A., & Prinz, P. (2022). The role of dietary sugars, overweight, and obesity in type 2 diabetes mellitus: a narrative review. *European Journal of Clinical Nutrition, 76*(11), 1497–1501. https://doi.org/10.1038/s41430-022-01114-5.

Witek, K., Wydra, K., & Filip, M. (2022). A high-sugar diet consumption, metabolism and health impacts with a focus on the development of substance use disorder: A narrative review. *Nutrients, 14*(14), 2940. https://doi.org/10.3390/nu14142940.

Yablonski-Alter, E., Agovic, M., Gashi, E., Lidsky, T., Freedman, E., & Banerjee, S. (2008). Chronic cocaine enhances release of neuroprotective amino acid taurine: a microdialysis study. *Nature Precedings*. https://doi.org/10.1038/npre.2008.2319.1.

Yudkin, J. (1972). *Pure, White and Deadly*. Viking Press.

Zink, J. I., Hardy, G. E., & Sutton, J. E. (1976). Triboluminescence of sugars. *The Journal of Physical Chemistry, 80*(3), 248–249. https://doi.org/10.1021/j100544a007.

Chapter 2

Ethanol

> "I have taken more out of alcohol than alcohol has taken out of me."
> Winston Churchill

The History of Ethanol

Ethanol can be considered a derivative of sugar due to its method of production, but it possesses a unique history and identity. Ethanol is a compound that has been present since the age of the most ancient civilizations. Unfortunately, the origin of the first distilleries and fermenters is unknown; fortunately, archaeological evidence suggests that alcoholic beverages have accompanied humans for a long time.

The word alcohol is often used, and in some dictionaries and slang, it is synonymous with ethanol. Less-used terms to refer to ethanol are spirits of wine and grain alcohol. In scientific language, besides the name ethanol, it can be called ethyl alcohol, methyl carbinol, ethyl hydroxide, ethyl hydrate, and anhydrol. For convenience and unless otherwise indicated, ethanol and alcohol are used interchangeably in this chapter.

The word alcohol derives from the Arabic-Hispanic *al-kuḥúl* and the classical Arabic *al-kuḥl*. The use of the term alcohol spread during the medieval period, but before that, it had a different meaning. The Latin word *alcohol* originally referred to *galena*, a stibnite powder used as makeup in the past (Castellano Actual, 2015). Curiously, stibnite is not related to alcohol; stibnite is the mineral antimony(III) sulfide (Sb_2S_3), which, when burned in a flame, produces a white powder called antimony(IV) oxide (Sb_2O_4). In the past, people began using the word *alcohol* to refer to any compound obtained through sublimation and, by extension, to ethanol, since distillers purify this liquid through evaporation and condensation. Thus, the meaning of the word *alcohol* evolved to refer exclusively to the liquid rather than the method. This word has since been adapted into several languages, including English (alcohol), Welsh (alcohol), Irish (alcól), Greek (αλκοόλ), Spanish (alcohol), French (alcool), Portuguese (álcool), German (alkohol), Polish (alkohol), Danish (alkohol), Dutch (alcohol), Italian (alcol), Romanian(alcool), Russian(алкоголь), Turkish (alkol), Hungarian (alkohol), Swedish (alkohol), and Norwegian (alkohol), among others.

Most likely, alcoholic ferments made of cereals or fruits were the first preparations for nutritional and recreational purposes. Wine consumption seems to date back to prehistoric times. Archaeologists found grape remains in ceramic vessels in Zagros (now Iran) dated to 5400 BC. Also, in the Caucasus Mountains (between Armenia and Georgia), grape pips dating to 6000 BC were found, which could indicate a possible origin of wine (Charters, 2006).

In Asia, clay tablets describing cereal ferments, comparable to modern brews, were found in Sumerian settlements dated to 4000 BC; nonetheless, the fermentation process may have been known as far back as 10,000 BC, when the Proto-Sumerians were still nomadic groups (Sewell, 2014). However, specialists currently debate the dates, as they differ by thousands of years. Whereas in China, archaeologists have also found ancient microbotanical evidence of fermented cereals in ceramic vessels dated to the Neolithic period (9000–3800 BC), similar discoveries have been made in other regions in China. Additionally, samples of brewing from the Bronze Age (1900 BC) and bronze vessels containing organic residues from the Han Dynasty (200 BC–220 AD) have been discovered (Liu et al., 2024).

Native Americans independently developed fermentation methods using local plants. For example, the ancient Otomi culture in Mexico created a fermented beverage from agave, called pulque, around 2000 BC (Escalante et al., 2016). The historian Fray Bernardino de Sahagún (1499–1590) confirmed this by describing how natives produced and consumed pulque. Centuries later, this beverage is still produced for domestic consumption (Álvarez-Ríos et al., 2022).

Distant civilizations across different periods of history were able to produce both fermented and distilled beverages. Regarding distillation, it is likely that the technology for distilling alcohol was rediscovered multiple times throughout history. The indirect collaboration across cultures to perfect the art of alcohol distillation is particularly captivating. The earliest documented reference to distillation methods originates from the Akkadian Empire around 1200 BC, which describes perfumery and implies knowledge of distilling ethanol and other organic compounds in ancient Mesopotamia. Approximately a thousand years later, during the 2nd century BC, written records from China mention the production of the alcoholic beverage baijiu (Zheng & Han, 2016). However, cereal distillates were probably being produced even earlier in this region.

In Europe, Aristotle (384–322 BC) described the distillation process and noted the flammability of alcohol. He also commented on the dangers of alcohol consumption during pregnancy, stating: "Foolish and drunken and harebrained women most often bring forth children like unto themselves, morose, and languid." (del Campo & Jones, 2017).

It is unknown when research on ethyl distillates began in the Middle East, but part of the etymology of alcohol has an Arabic influence, possibly adopted during the 7th and 8th centuries. For example, the Greek word *ambix* (cup) was formerly used to refer to the instrument employed for the separation and purification of liquids. At some point, the Arabic prefix *al-* was added, forming the new words *al-anbîq*, *alembicus*, and finally alembic. In addition, the Baghdad bibliographer Ibn al-Nadīm (c. 932–995) wrote about the alchemist Jabir ibn Hayyan (c. 721–c. 806–816) from Tus (now in Iran), who was renowned for his studies in chemistry and medicine related to alcohol. By the early 10th century, Arabic medicine and its advanced surgical procedures had spread throughout Europe. Among other topics, alcohol was mentioned in the encyclopedia *Kitāb al-Taṣrīf*, known in English as *The Method of Medicine*, a classic work by the scientist al-Zahrāwī (c. 936–c. 1013), also known as Abulcasis, who was born in Medina Azahara, now part of Córdoba, Spain (Abdulrazeq et al., 2024).

The history of distillates played a central role in medieval Europe. By the 12th century, numerous recipes for producing ethyl distillates had been documented. New names alluding to water were assigned to alcoholic beverages, perhaps because they are also clear liquids—for example, *aqua ardens* ("burning water") and *aqua vitae* ("water of life") (Rasmussen, 2019). Another related term that emerged was 'liqueur,' which derives from the Latin *liquor* (liquid). This demonstrates the popularity of alcoholic drinks, as reflected in the many related terms.

Meanwhile, ethanol was studied for medical applications. For example, Tadeo Alderotti (1223–1296) described a method for obtaining 90% pure ethanol through fractional distillation. European physicians Arnau de Vilanova (c. 1240–c. 1311) and Jean de Roquetaillade (also known as John of Rupescissa) (c. 1310–c. 1366–1370) mentioned alcohol in their works, referring to it as the "quintessence of wine." They observed that the use of high-purity alcohol improved patients' chances of recovery, even though microorganisms had not yet been discovered.

The high consumption of alcoholic beverages among the population was an important source of income for governments during the medieval era. This led to establishing tests to determine the alcohol content of beverages and to impose a fairer tax. In 16th-century England, one such test involved dropping gunpowder into a flame of the analyzed alcoholic beverage. The tax rate was adjusted depending on whether the gunpowder ignited or not. This practice is believed to be the possible origin of the unit "proof," which refers to this test, trial, or demonstration (Jensen, 2004). However, it was not until 1816 that the English scale was standardized based on the specific gravity of alcohol, ranging from 0 to 175 proof, although this scale is now obsolete.

During the 19th century, ethanol measurement was refined, notably in 1824 by the French chemist Louis Joseph Gay-Lussac (1778–1850), who developed an instrument based on the difference in density between ethanol and water. The Gay-Lussac scale, which measures alcohol by volume (ABV), ranges from 0 to 100%, assigning 100% ABV to pure alcohol and 0% ABV to pure water (Crosland, 1978). Meanwhile, in 1848, the United States developed its own system, adopting the proof unit, which arbitrarily assigned 100 proof to a concentration of 50% ABV—a standard concentration for spirits at the time. Consequently, in the U.S. proof system, 200 proof corresponds to pure alcohol, while 0 proof corresponds to pure water.

Society of the 19th century witnessed significant advances related to ethanol, as this period saw major contributions to science. In 1826, the English chemist Henry Hennell (c. 1797–1842) mentioned in his article that his compatriot, Michael Faraday (1791–1867), had described the synthesis of ethanol by the hydration of ethylene gas in aqueous sulfuric acid years earlier (Hennell, 1828). This discovery served as a precedent for the method later used by the Union Carbide Company to produce synthetic ethanol via the acid-catalyzed hydration of ethylene gas (Union Carbide Corporation, 2024). Although ethanol synthesis was not particularly significant in the 19th century, this changed in subsequent years, as the demand for ethanol increased due to the industrialization of the world in the early decades of the 20th century and the growing social demand for alternative fuels. Furthermore, the development of the petroleum derivatives industry facilitated the production of synthetic ethanol from natural gas through a profitable process ("Synthetic Ethanol Capacity on Rise," 1966).

Also in the 19th century, during the heyday of chemistry, when new compounds were being discovered and synthesized, the need for a more formal classification arose. In 1834, Jean Baptiste Dumas (1800–1884) and Eugène Peligot (1811–1890) isolated methanol from wood spirit (Wisniak, 2009). This newly identified compound shared key properties with ethanol—until then known simply as alcohol—including reactivity and structural features such as an alkyl radical and a hydroxyl group. Recognizing the need for a clearer classification, the Swedish chemist Jöns Jacob Berzelius (1779–1848) proposed the general term "alcohol" for such compounds. He assigned the name "methanol" to wood alcohol and "ethanol" to wine alcohol (Rasmussen, 2019). This nomenclature was adopted by the scientific community, and the term alcohol came to refer to all compounds with the OH functional group. Nonetheless, due to historical associations, the terms alcohol and ethanol are still used interchangeably in some social contexts.

The history of ethanol is vast and can be divided into two major aspects: the first pertains to its role in alcoholic beverages, while the second encompasses its diverse applications in industry, medicine, and fuel production. Over time, various uses for ethanol have emerged, and other uses have been preserved for centuries. For example, in medicine, ethanol has been used as an antiseptic and disinfectant. In herbal medicine, it has served as a solvent for organic extracts and concoctions, while in perfumery, it is still used as a solvent and a preservative. Moreover, in the Modern and Contemporary Ages, solvents such as ethanol have played a crucial role in accelerating the findings in chemistry (Mainkar et al., 2024).

General Data on Ethanol

Ethanol is a colorless and water-miscible liquid. It has a characteristic odor and a spicy taste. Some also sense a bitter or sweet flavor due to trace compounds (Scinska et al., 2000). Ethanol (C_2H_6O) is an organic compound that belongs to the alcohol family and is the second simplest after methanol (CH_4O). Alcohols have the general formula ROH, where R is an aromatic or aliphatic fragment (ethanol belongs to the aliphatic alcohols).

Ethanol has a molecular weight of 46.07 g/mol; the hydroxyl group (-OH) is attached to the linear ethyl chain fragment, so the formula is CH_3CH_2OH. Ethanol has a boiling point of 78.4°C, melting point of -114.1°C, density of 0.79 g/ml, and flash point of 13.0°C (Merk, 2024). Ethanol is a volatile liquid. At 25°C, its vapor pressure is about 58.7 mm Hg, over twice that of water, about 23.8 mm Hg.

Conventionally, alcohol by volume (alc./vol. or ABV) is defined as the number of milliliters (mL) of pure ethanol per 100 mL of solution. The ethanol content in alcoholic beverages is measured with instruments such as a hydrometer (or densitometer), vinometer, ebulliometer, and refractometer (Jaywant et al., 2022). Ethanol (CAS number 64-17-5) is commonly available as a reagent or solvent at a concentration of 96 ABV, which consists of 96% ethanol and 4% water. It is available in varying purity levels, ranging from technical grade to anhydrous and HPLC.

According to SDS, ethanol is flammable. So, storing it at 2-30°C is recommended. According to the NFPA 704 hazard identification system, ethanol is classified as hazardous, indicating moderate health risk, high flammability, and stable reactivity. Health = 2 (blue); very flammable = 3 (red); stable, reactivity = 0 (yellow); and no indication is given for special risks (white).

OH	**ETHANOL**	
	Other names: Alcohol, spirits of wine, grain alcohol, ethyl alcohol, ethyl hydroxide, ethyl hydrate, and anhydrol	
	Origin: Natural/synthetic	Formula: C_2H_6O
Type: Aliphatic alcohol	Discoverer: Unknown	MW: 46.07 g/mol
Appearance: Transparent liquid	Inventor: NA	mp: -114.1°C
Uses: Beverage, psychotropic, antiseptic, disinfectant, sanitizer, solvent, and combustible	Toxicity: Yes	bp: 78.4°C

If ethanol is accidentally ingested, rinse the mouth with water, but only if the person is conscious. In case of accidental ingestion, rinse the mouth and drink water, only if the person is conscious; otherwise, do not give any liquid and do not induce vomiting. If poisoning is suspected, monitor the victim and get medical help. Monitor for signs such as decreased heart rate, low blood pressure, low temperature, sedation, slow breathing, loss of balance, restlessness, agitation, and slurred speech. Some symptoms may also occur, such as relaxation, euphoria, dizziness, fatigue, nausea, and blackout (Adger & Saha, 2013).

The Structure and Reactivity of Ethanol

Ethanol is a linear aliphatic alcohol; carbons and oxygen atoms have sp^3 hybridization, forming σ covalent bonds with hydrogen and oxygen, with a dipole moment of μ_{exp} = 1.69 D (Jorge et al., 2022). The crystal structure, determined by XRD at low temperature (-186°C), indicates that neighboring ethanol molecules adopt *gauche* and *trans* conformations, which form intermolecular hydrogen bonds with lengths of 2.716 Å and 2.730 Å. The cell parameters are monoclinic, with space group *Pc*, and dimensions a = 5.377 Å, b = 6.882 Å, and c = 8.255 Å (Jönsson, 1976).

The spectroscopic characterization of ethanol is easy because it is a small molecule. The ^{1}H-NMR in $CDCl_3$ (Spectral Database for Organic Compounds, 2024) shows three signals: the CH_3 signal at 1.23 ppm, the CH_2 signal at 3.69 ppm, and the OH signal at 2.61 ppm. The ^{13}C-NMR in $CDCl_3$ shows two signals: CH_3 at 18.13 ppm and CH_2 at 57.79 ppm. The IR spectrum in the liquid phase shows a characteristic OH stretching band of alcohols at 3358 cm^{-1}, a set of aliphatic CH stretching bands in the range of 2974–2927 cm^{-1}, and confirmation of the C–O bond at 1050 cm^{-1}.

Ethanol is completely miscible in water, forming an azeotrope. This means that mixtures of water and ethanol behave as a single liquid, making their separation more difficult. Also, ethanol easily absorbs water from the air. The mixture is strong because of vapor pressure and intermolecular H-bonding interactions (Carravetta et al., 2022). Ethanol is dried using standard methods, such as distillation with iodine-activated magnesium (Mg/I_2), potassium hydroxide (KOH), or calcium oxide (CaO). A nitrogen atmosphere and molecular sieves help maintain the dryness of ethanol (Williams & Lawton, 2010).

Ethanol and mixtures with water are used in natural product extraction processes and as diluents in formulations for perfumery, pharmacy, synthesis, or processed foods. The biocompatibility of ethanol is one of the main reasons for its use. Furthermore, from the perspective of green chemistry, ethanol, water, and their mixtures—being oxygenated and polar solvents—are considered green, providing an additional reason to prefer ethanol over other organic solvents (Capello et al., 2007).

Acetic acid, acetaldehyde, ethyl ether, chloroethane, and ethyl acetate are organic compounds derived from ethanol (Figure 3). They are widely used as solvents and reagents in synthesis at different scales for multiple purposes.

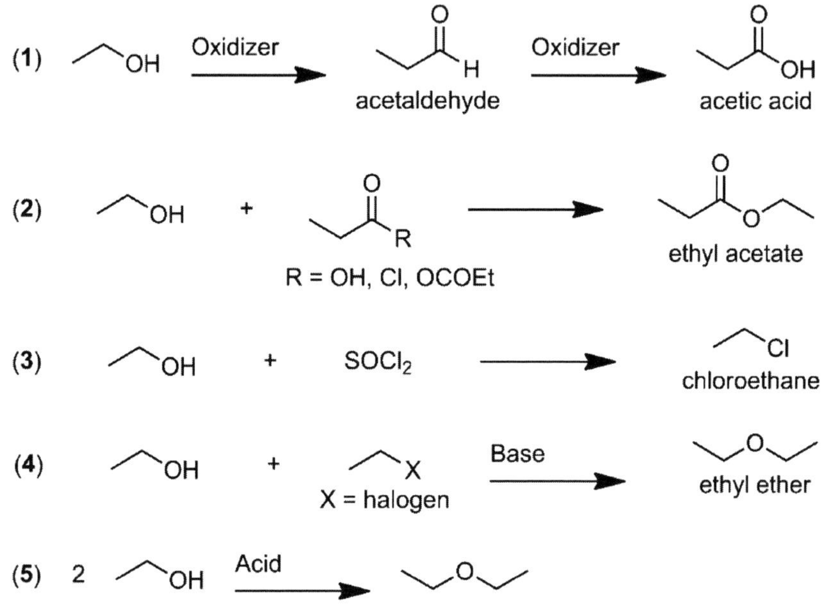

(1) OH —Oxidizer→ acetaldehyde —Oxidizer→ acetic acid

(2) OH + R (R = OH, Cl, OCOEt) → ethyl acetate

(3) OH + $SOCl_2$ → chloroethane

(4) OH + X (X = halogen) —Base→ ethyl ether

(5) 2 OH —Acid→ O

Figure 3. Reaction schemes for the synthesis of different solvents and reagents from ethanol.

The reactivity of ethanol corresponds to that of a primary alcohol. It undergoes substitution, oxidation, and condensation reactions. The following are some examples (Wade, 2017; Warren & Wyatt, 2004). Primary alcohols are oxidized with Collins reagent, a mixture of chromium trioxide (CrO_3), pyridine, and dichloromethane, to form aldehydes (RCHO). Pyridinium chlorochromate (PCC), in an acidic medium can also oxidize alcohols to obtain the corresponding RCHO. When the primary alcohol reacts with a stronger oxidizing agent, such as potassium dichromate ($K_2Cr_2O_7$) or potassium permanganate ($KMnO_4$) in an acidic medium, the corresponding carboxylic acid (RCOOH) forms. Primary alcohols are halogenated using thionyl chloride ($SOCl_2$) or phosphorus pentachloride (PCl_5), and hydrobromic acid (HBr), phosphorus tribromide (PBr_3), or bromine and red phosphorus (P_8/Br_2) to form the corresponding chlorinated (RCl) or brominated derivatives (RBr), respectively. However, the efficient synthesis of the iodinated compound requires an additional step, in which the bromoalkane is first obtained and subsequently reacts with NaI in a nucleophilic substitution (Finkelstein reaction). Alcohols participate in esterification reactions (Fischer-Speier esterification), where the alcohol is condensed with RCOOH, an anhydride (RCOOCOR), or an acyl chloride (RCOCl). Also, primary alcohols react with alkyl halides (Williamson ether synthesis) in an alkaline medium to form ethers (ROR') via a nucleophilic substitution mechanism. Symmetric ethers can also be synthesized by condensing two alcohols via an acid-catalyzed nucleophilic substitution mechanism.

Qualitative tests can be used to identify alcohols. For example, the Lucas test differentiates primary alcohols from secondary and tertiary alcohols. In this test, an aqueous acidic solution of $ZnCl_2$ becomes cloudy in the presence of secondary or tertiary alcohols due to the rapid formation of the corresponding halide. In contrast, primary alcohols, such as ethanol, do not react at room temperature.

Oxidation with PCC is another method to distinguish primary alcohols from secondary and tertiary alcohols. Primary alcohols, like ethanol, are oxidized to an aldehyde. The resulting aldehyde reacts with Tollens' reagent, $[Ag(NH_3)_2]^+$, to form a silver mirror by reducing silver(I) ions to metallic silver.

Metabolism and Toxicological Effects of Ethanol

Ethanol acts as both a stimulant and a depressant of the central nervous system. In the brain, ethanol enhances dopamine activity, contributing to an initial feeling of well-being due to its rapid caloric availability. However, it suppresses glutamatergic neuronal activity, leading to anxiety and discomfort (Wilson & Matschinsky, 2020).

Most of the ingested alcohol is absorbed in the stomach and small intestine. Approximately 90% of ethanol is metabolized by the liver, while the remaining 10% is eliminated through the kidneys and lungs (Pietraszek et al., 2010). In the liver, ethanol metabolism begins with the enzyme alcohol dehydrogenase (ADH), which catalyzes the oxidation of ethanol to produce acetaldehyde. Subsequently, the enzyme aldehyde dehydrogenase (ALDH) and the coenzyme nicotinamide adenine dinucleotide (NAD) catalyze the conversion of acetaldehyde into acetate, with the concomitant production of reduced nicotinamide adenine dinucleotide (NADH) as a byproduct of ethanol oxidation (Figure 4).

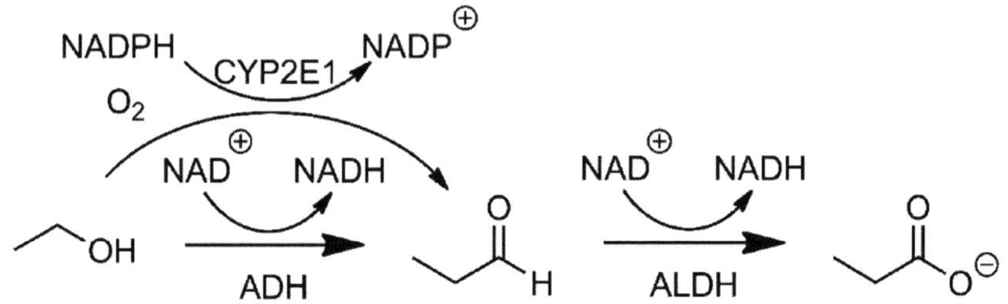

Figure 4. A simplified pathway of ethanol metabolism, mediated by the enzymes ADH and ALDH, along with the oxidation products acetaldehyde and acetate.

NAD/NADH imbalance, resulting from chronic alcohol consumption, interferes with the function of various biomolecules, including carbohydrates, lipids, proteins, and certain steroid hormones. For example, redox imbalance leads to lactic acidosis and hyperuricemia, promoting fatty acid oxidation, which, in turn, favors triglyceride production—a process associated with hepatic steatosis (Subramaniyan et al., 2021). Additionally, acetaldehyde impairs protein function and contributes to liver damage. A fatty liver is dysfunctional, affecting the metabolism of lipids, carbohydrates, proteins, and purines (Wilson & Matschinsky, 2020). The progression of these lesions and dysfunctions ultimately leads to more severe conditions, such as liver cirrhosis (Narro et al., 2024). Alcohol-induced liver injury must be confirmed through liver biopsy, as the condition may initially be asymptomatic (Sorbi et al., 2000).

Ethanol metabolism produces various effects that impact individuals differently. Several studies suggest that alcohol consumption affects women more than men on average (Thomasson, 2002). One contributing factor is the lower activity of the enzyme chi-ADH during first-pass metabolism in women. Additionally, gastric emptying with alcohol is 42% slower in women, while hepatic oxidation is 10% higher, exposing them to higher concentrations of acetaldehyde (Baraona et al., 2001). Another factor is body weight, as men are, on average, taller and heavier than women. Consequently, with equal alcohol consumption, women exhibit a higher blood alcohol concentration than men (McCaul et al., 2019).

Furthermore, animal studies have shown that long-term ethanol consumption negatively impacts oocyte quality in female mice. Findings indicate that ethanol concentrations exceeding 1% lead to an increase in reactive oxygen species (ROS), which impair oocyte maturation (Kizasu et al., 2024).

Finally, repeated alcohol consumption during pregnancy has detrimental effects on offspring. It increases the risk of fetal alcohol syndrome, which is characterized by poor growth (Ornoy & Ergaz, 2010). Moreover, it contributes to fetal alcohol spectrum disorders, which manifest through various features, including small head circumference, structural defects of the central nervous system, short palpebral fissures, smooth philtrum, thin upper lips, long philtrum, short anteverted nose, ptosis of the eyelids, epicanthal folds, strabismus, midface hypoplasia, railroad track-like ears, and abnormal palmar creases (del Campo & Jones, 2017).

Ethanol resistance varies significantly among individuals of the same sex due to genetic factors (Edenberg, 2007). The enzymes alcohol dehydrogenase (ADH) and aldehyde dehydrogenase (ALDH) play a central role in ethanol metabolism. Both enzymes are encoded by distinct genes with multiple variants (also known as alleles), some of which influence alcohol consumption and the risk of developing alcoholism. For instance, variants of the *ADH1B*, *ADH1C*, and *ALDH2* genes produce enzymes with altered kinetic properties. The *ADH1B* and *ADH1C* variants encode ADH enzymes that metabolize ethanol to acetaldehyde more rapidly, even at low alcohol concentrations. In contrast, the *ALDH2* gene encodes a virtually inactive ALDH enzyme, leading to acetaldehyde accumulation. In all three cases, these genetic variants appear to confer a protective effect against alcoholism (Zaso et al., 2019). Conversely, the *ADH4* variant encodes an ADH enzyme that significantly contributes to ethanol oxidation at high concentrations, predisposing individuals to higher alcohol consumption and an increased risk of alcoholism (Luo et al., 2006). Regarding the relationship between ethnic origin and the analyzed genes, studies have shown that the distribution of the *ADH1B* and *ALDH2* variants is higher in East Asian populations, whereas *ADH4* variant is more prevalent in populations of European and American descent, increasing their susceptibility to alcoholism.

Age is a crucial factor in ethanol consumption. Studies indicate that initiating alcohol consumption during adolescence increases the likelihood of continued use in adulthood and raises the risk of developing addiction and related diseases. Consequently, it is recommended to delay the onset of alcohol consumption until adulthood to reduce these risks (Moure-Rodríguez & Caamano-Isorna, 2020). Furthermore, research shows that alcohol consumption harms the nervous system regardless of age, affecting both adolescents and adults (Pérez-García et al., 2022). In addition to neurological effects, alcohol use poses several other health risks.

There are various strategies to reduce alcohol consumption. One such strategy is voluntary abstinence; however, this approach may be challenging for some individuals due to withdrawal-related side effects. Pharmacological interventions represent another option. For instance, animal studies suggest that the administration of the antagonist drug S1RA, a modified morpholine (CAS number 878141-96-9), can significantly reduce ethanol intake in adulthood and even induce an aversion to alcohol consumption (Salguero et al., 2024). However, the use of S1RA raises ethical concerns, as it involves administering a drug with unknown long-term side effects to counteract the consumption of a substance with well-documented short-, medium-, and long-term consequences.

For this reason, some researchers propose the use of cannabis as a recreational drug to reduce ethanol consumption, as it contains tetrahydrocannabinol (THC), a psychoactive that affects the central nervous system and induces feelings of well-being and relaxation (Reiman, 2009). However, the use of medicinal cannabis as an alternative treatment for alcohol addiction remains a subject of debate due to several limitations.

The first limitation of cannabis pertains to health, as further research is required to determine the side effects associated with prolonged consumption. The second limitation is the legal aspect, as recreational cannabis use remains prohibited in most countries. Additionally, many jurisdictions restrict the medical use of cannabidiol (CBD), despite its demonstrated efficacy as a muscle relaxant and seizure-reducing agent in patients with Lennox-Gastaut and Dravet syndromes (Berg et al., 2024).

Therefore, further studies are needed to determine whether cannabis-based therapies can aid in alcohol addiction rehabilitation. The ongoing debate primarily concerns the potential long-term effects of cannabis use, particularly those related to THC (Zehra et al., 2018) or CBD (Kessler et al., 2021).

As a final note, it is important to recognize that, regardless of gender, age, or genotype, no one is exempt from the risks associated with alcohol consumption.

Obtaining Alcoholic Beverages and Ethanol

Apart from beer, the most common sources of ethanol for the average consumer are distilled spirits, such as vodka, tequila, rum, brandy, and whisky, among others. Each type of spirit contains small amounts of higher alcohols and esters that contribute to its characteristic taste and aroma; however, ethanol remains the primary component in all cases. Despite technological advancements over time, the fundamental logistics of ethanol distillation remain consistent, regardless of production volume or brand. The following section explores the biosynthesis and distillation of ethanol and alcoholic beverages. Additionally, a method for synthesizing ethanol from natural gas is discussed.

Biosynthesis and Distillation of Ethanol

The primary method for obtaining ethanol is through the fermentation of simple sugars by microorganisms. Certain yeasts and bacteria, such as the single-celled fungus *Saccharomyces cerevisiae* and the bacterium *Zymomonas mobilis*, efficiently facilitate ethanol biosynthesis. Fermentation occurs when microorganisms cannot respire due to the absence of oxygen, leading them to generate energy via the partial oxidation of polysaccharides, disaccharides, and glucose (Ciani et al., 2013). During glycolysis, pyruvate, ATP, and NADH are synthesized, among other molecules. Pyruvate is then decarboxylated to acetaldehyde by the enzyme pyruvate decarboxylase. Finally, NADH reduces acetaldehyde to ethanol as a by-product, with ATP serving as a chemical energy reservoir.

Ethanol distillation for human consumption involves multiple stages, beginning with the selection of sugar-rich ingredients. Sugarcane is used to produce rum and cane alcohol; barley, wheat, rye, and corn are used to produce whiskey; agave is used for tequila and mezcal; rice is used for sake; and vodka can be produced from potatoes, grapes, or any fermentable raw material, including the aforementioned ingredients. In the next stage, the selected raw materials are harvested and transported. They are then crushed, cut, ground, or macerated at the production facility.

Following this process, the must is prepared by adding water if necessary and heating it to facilitate the release of sugars from the biomass. Once cooled to 30–35°C, yeast reproduction is promoted, and additional yeast may also be inoculated. During this fermentation process, microorganisms consume sugars to produce ethanol, along with small quantities of other alcohols, such as polyphenols, which have been characterized in the final product (e.g., quercetin, fisetin, para-hydroxyphenylacetate, 2-phenylethanol, resveratrol, *para*-coumaric acid, and sinapic acid) (Chrzanowski, 2020).

Once the ferment reaches a specific alcohol content, it is transferred to a distillation apparatus, typically made of copper or stainless steel. Distillation concentrates the alcohol through heating, evaporation, and condensation (Marisa Mendes, 2017). Methanol residues are found in the head fraction of the distillate (boiling point 64.7°C), while the heart fraction contains most of the ethanol (boiling point 78.4°C). The tail fraction consists of water and higher alcohols, such as polyphenols and polyols. This distillation process may be repeated to increase the concentration and purity of ethanol. Double and triple distillations are common in the production of distilled alcoholic beverages.

Aging is an optional step, typically applied to a portion of production (Silvello et al., 2021). During aging, the liquid matures in wooden or glass barrels, allowing for the formation of esters that enhance aroma and other compounds that contribute to the beverage's color and flavor. Before bottling, the ethanol is diluted with distilled water to ensure it is safe for consumption.

Beer is a brewed beverage made from a blend of malted barley, water, and hops—the ingredient that gives beer its characteristic bitterness. The production process shares several steps with distilled beverages until fermentation; however, unlike distilled spirits, beer is not distilled afterward. Instead, the fermented liquid goes through additional steps: it is cooled to 0°C to allow yeast and solids to settle, matured to develop flavors, filtered to remove any remaining residues, and finally bottled (Ingledew & Hysert, 1994).

Synthetic Ethanol

Currently, there are several methods for synthesizing ethanol in the laboratory; however, many of these options are not viable due to low yields, poor atomic economy, or high costs. For example, the reduction of acetaldehyde or acetic acid with hydrides yields ethanol as the product; however, these reagents are not suitable, as carbonyl compounds are typically more valuable and allocated for other processes.

Therefore, this section focuses on a method used on an industrial scale, which was developed more than a century ago: the synthesis of ethanol via the hydration of ethylene with aqueous sulfuric acid under high temperature and pressure (Nelson et al., 2009). Notably, ethanol can also be dehydrated in an acidic medium to yield ethylene as an elimination product (Figure 5.1). The key to achieving a high conversion rate lies in the use of metal oxides, molecular sieves, or a Lewis acid catalyst (Zhang & Yu, 2013).

The complete synthesis of ethanol from hydrocarbons occurs in only two stages (Figure 5.2). First, ethane and propane from natural gas undergo high-pressure, high-temperature processing to produce ethylene gas (Union Carbide Corporation, 2024). The ethylene is then transferred to a separate reactor, where the gas undergoes hydration in an aqueous medium with sulfuric acid as a catalyst to produce ethanol (Britton, 1969).

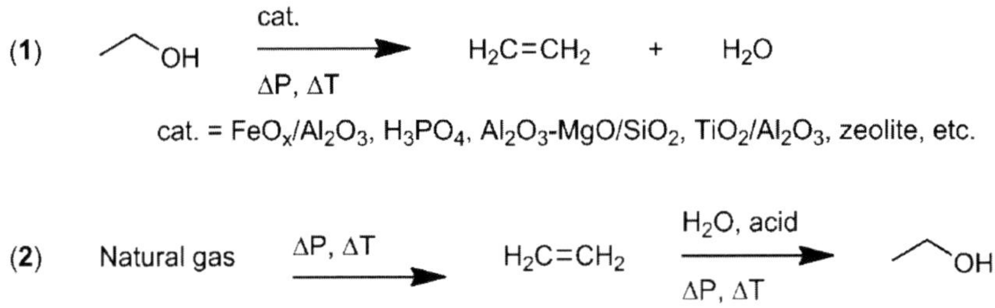

(**1**) $CH_3CH_2OH \xrightarrow[\Delta P,\ \Delta T]{\text{cat.}} H_2C{=}CH_2 + H_2O$

cat. = FeO_x/Al_2O_3, H_3PO_4, Al_2O_3-MgO/SiO_2, TiO_2/Al_2O_3, zeolite, etc.

(**2**) Natural gas $\xrightarrow{\Delta P,\ \Delta T} H_2C{=}CH_2 \xrightarrow[\Delta P,\ \Delta T]{H_2O,\ \text{acid}} CH_3CH_2OH$

Figure 5. Synthesis of ethylene from ethanol (1) and synthesis of ethanol from natural gas (2).

Bioethanol Fuels

Bioethanol is ethanol produced naturally through biosynthesis. Bioethanol fuels are combustible fuels that contain a proportion of bioethanol. In general, synthetic ethanol and bioethanol, as components of fuel blends, serve as alternative fuels, and their expanded use is feasible for several reasons discussed below.

First, rising fossil fuel prices have increased the demand for alternative energy sources, particularly for transportation. Currently, the costs of gasoline and diesel continue to rise. In addition, potential shortages and market speculation by certain countries within the Organization of the Petroleum Exporting Countries (OPEC) create significant uncertainty (OPEC, 2024).

Consequently, alternative fuels are becoming viable options in various geographical regions.

Sustainable bioethanol fuel can be produced from low-impact crops. For example, lignocellulose, a low-value waste product, can be utilized (Hahn-Hägerdal et al., 2006). In contrast, bioethanol derived from sugarcane—also used as a raw material for sucrose production—may contribute to price increases for both products. This issue arises when biofuels and food production compete for sugarcane crops, especially during periods of crisis.

Ethanol serves as an effective fuel additive when blended with regular gasoline; however, before blending, water must be removed from ethanol to achieve a purity of at least 99.5%. Once dried, ethanol can be mixed with gasoline to create gasohol, a fuel blend. The E10 blend, which contains 10% ethanol by volume, closely resembles regular gasoline in performance (Murachman et al., 2014). Other fuel blends with higher ethanol content, such as E75 and E85, contain 75% and 85% ethanol, respectively. In the United States, flex-fuel vehicles can currently run on ethanol-gasoline blends ranging from E10 to E85 (Pouliot et al., 2018).

Despite the advantages of ethanol-blended fuels, certain emission-related challenges persist. While nitrogen oxide (NOx) emissions decrease, emissions of carbon monoxide, methane, formaldehyde, acetaldehyde, and ethanol increase (Suarez-Bertoa et al., 2015). Therefore, advancements in combustion systems are necessary to mitigate these emissions. On the positive side, some automobile vehicles are already equipped with engines adapted for biofuels from E10 to E85.

From an energy perspective, the autoignition temperature of pure ethanol is 368°C (Chen et al., 2010), with a heat of combustion of −1366.7 kJ/mol (Schmidt-Rohr, 2015). In comparison, octane has an autoignition temperature of 220°C and a heat of combustion of 5485 kJ/mol. Numerically, ethanol is inferior to octane in energy content; however, producing octane-based fuels (gasoline) is typically associated with significant environmental damage and contamination (Stikkers, 2002). For this reason, ethanol remains a viable alternative despite its not up-to-par energy parameters.

Ethanol Production and Consumption Worldwide

Many spirits and liqueurs result from processes developed and perfected over centuries and are protected by a designation of origin. These beverages must adhere to strict standards regarding cultivation, ingredients, aging, alcohol content, and packaging. Examples of origin-designated alcoholic beverages include:

- USA: Tennessee whiskey
- Mexico: Tequila, mezcal, bacanora, sotol, charanda, and raicilla
- Peru and Chile: Pisco
- Bolivia: Singani
- Brazil: Cachaça
- Cuba: Rum
- China: Baijiu, shaojiu, mijiu, and maotai
- Korea: Soju
- UK: Scotch and Irish whisky
- France: Cognac
- Spain: Pacharán and Jerez brandy
- Cyprus and Greece: Ouzo
- Italy: Grappa

In addition, various alcoholic beverages without a designation of origin, such as brandy, vodka, schnapps, and sake, are produced globally.

Alcoholic beverage production occurs worldwide, primarily in Asia, the Americas, and Europe (Institute of Alcohol Studies, 2020). The leading alcohol-producing countries in 2019 reported the following key data (Table 1):

1. China – The world's largest ethanol producer in 2019. With 9.6 million km^2 of land, China has significant agricultural potential. Its climate is well-suited for cultivating cereal crops such as sorghum, rice, and wheat.
2. United States – A country with a long tradition of alcoholic beverage production, including fermented beverages such as beer and wine, as well as distilled spirits such as vodka, whiskey, and rum.
3. Brazil – A country leading ethanol producer in South America and Latin America. Its extensive sugarcane plantations, cultivated in a warm, tropical climate, play a crucial role in ethanol production.
4. Mexico – Known for its wide variety of distilled spirits, including mezcal and tequila, which are major exports, as well as charanda and sotol for domestic consumption. Mexico also produces fermented

beverages, beer for domestic consumption and export, and pulque for local consumption.

5. Germany – The top producer of alcoholic beverages in Europe specializes in fermented products such as beer and wine, alongside a significant production of spirits.
6. Russia – The second-largest alcohol producer in Europe and, until 2019, was the largest exporter of vodka. Most of Russia's alcohol production is intended for domestic consumption.
7. France – A leading European producer, particularly renowned for wines and spirits with a designation of origin, such as the world-famous cognac and champagne.
8. Spain – A country with a rich history of alcoholic beverage production. Spain produces a wide range of fermented drinks, including cider and wine, as well as distilled spirits such as grape marc, sherry brandy, and anise liqueur.
9. Italy – A competitive producer of beer and spirits, best known for its globally recognized wines, such as chianti, lambrusco, and prosecco.
10. India – A nation that has a long-standing tradition of producing alcoholic beverages, with sugarcane distillates being the most prominent.

Table 1. Leading countries in alcoholic beverage production in 2019 (Insider Monkey, 2024)

Country	Production (thousand metric tons)
China	55,112
USA	26,598
Brazil	17,076
Mexico	12,542
Germany	9,984
Russia	9,160
France	8,981
Spain	8,299
Italy	7,350
India	6,885

The most valuable alcoholic beverage companies originate from private industrial associations, excluding Chinese companies with state involvement. The most valuable companies, along with their respective brands (in parentheses), are Kweichow Moutai Co. Ltd. (*Moutai*), Anheuser-Busch

InBev (*Corona, Budweiser, Modelo*), Heineken International (*Heineken*), Diageo (*Johnnie Walker*), Wuliangye Yibin Co. (*Wuliangye*), Brown-Forman Co. (*Jack Daniel's*), James Hennessy & Co. (*Hennessy*), and Luzhou Laojiao Co. (*Guojiao National Cellar 1573*) (Table 2).

Table 2. The world's top 10 most valuable alcohol brands in 2024 (Statista, 2024)

Brand	Value (million dollars)
Kweichow Moutai	85,565
Corona	19,043
Budweiser	13,773
Heineken	12,821
Modelo	11,353
Johnnie Walker	10,545
Wuliangye	9,146
Jack Daniel's	7,121
Hennessy	7,095
National Cellar 1573	7,027

In 2024, the most valuable alcohol brand was a baijiu liquor produced in China. Notably, the top three positions were occupied by baijiu liquors, primarily consumed domestically. Interestingly, beer brands ranked second through fifth, highlighting the global popularity of beer. The ranking was completed by one Scotch whisky, one American whiskey, and one cognac.

The following provides an overview of the top-ranked brands:

1. Moutai, a baijiu liquor produced in China from a blend of cereals, has been manufactured since 1951.
2. Corona, a pale lager beer, has been produced in Mexico since 1926.
3. Budweiser, a lager beer, has been brewed in the United States since 1876.
4. Heineken, a pilsner beer of Dutch origin, has been produced since 1873.
5. Modelo, another Mexican beer, has been manufactured since 1925.
6. Johnnie Walker, a Scotch whisky, has been produced in Scotland since 1820.
7. Wuliangye, a Chinese baijiu liquor, has been in production since 1952.

8. Jack Daniel's, an American whiskey, has been distilled in the United States since 1875.
9. Hennessy, a French cognac, has been produced since 1765.
10. Guojiao National Cellar 1573, the oldest brand on the list, is a Chinese baijiu liquor produced since 1573.

Data on alcohol consumption by country and gender provides insights into who the highest consumers are and where they are located (**Table 3**). The amount of pure ethanol consumed by individuals over the age of 15 is significantly higher among men than women, often doubling or even tripling in comparison. In 2019, Romania ranked first in alcohol consumption per capita.

Although alcoholic beverages are consumed worldwide, Europe has the highest levels of consumption. Notably, 8 of the top 10 alcohol-consuming countries are in Europe, with 6 of them in Eastern Europe. In contrast, no countries from the Americas or Asia are included in the top 10, considering Georgia as part of Europe.

Table 3. Ranking of alcohol consumption per person (aged 15 and older) in 2019, measured in liters of pure alcohol (Ritchie & Roser, 2024)

Country	Consumption per person (l)	Men's consumption (l)	Women's consumption (l)
Romania	16.99	27.30	7.51
Georgia	14.33	24.31	5.86
Czechia	13.29	21.09	5.84
Latvia	13.09	21.73	5.99
Germany	12.22	19.22	5.49
Uganda	12.21	19.93	4.88
Seychelles	11.99	18.41	4.72
Austria	11.97	18.81	5.46
Bulgaria	11.92	19.51	4.91
Lithuania	11.79	19.34	5.39

The Prohibition of Ethanol

What would happen if alcohol consumption were completely banned? This question is relatively easy to answer, as it has occurred both in the past and present. The case of Prohibition in the United States from 1920 to 1933 is well-documented. In the years leading up to 1920, a conservative movement composed of Protestants, Progressives, and women advocated for the prohibition of alcohol. Their efforts led Congress to ban the production, sale, and consumption of alcoholic beverages. As a result, alcohol consumption declined; however, illegal trafficking surged. In fact, some criminal organizations thrived by diversifying into other illicit activities, even after the revoke of Prohibition (Britannica, 2024).

Today, alcoholic beverages are legal and socially accepted in most parts of the world, with legal drinking ages typically set at 18 or 21 years, depending on the country. However, in some nations governed by Islamic law (Sharia), alcohol is banned or heavily restricted. Most of these countries are in the Middle East and Africa. As of 2024, the list of countries with alcohol bans includes Afghanistan, Saudi Arabia, the United Arab Emirates, Libya, Mauritania, Kuwait, Yemen, Somalia, Brunei, Pakistan, Iran, Sudan, and the Maldives (Geeks for Geeks, 2024). The degree of enforcement varies; in some countries, alcohol consumption is tolerated in designated areas, often for tourism purposes, while in others, alcohol is strictly prohibited (World Health Organization, 2004).

For instance, in Afghanistan, following the Taliban's takeover in 2021, the consumption of alcoholic beverages was completely banned, with severe punishments for offenders. In contrast, the United Arab Emirates has designated tolerance zones for tourists and non-Muslim residents, who, with the appropriate permits, can purchase alcohol in licensed establishments such as restaurants and bars. However, the sale, distribution, and consumption of alcohol in public spaces remain prohibited, with violations subject to legal penalties.

If ethanol production for industrial purposes—such as sanitizing materials, cosmetics, and food additives—were to be halted, finding alternative solvents with comparable biocompatibility and functionality would be extremely challenging. Ethanol is unique in its versatility, and while substitutes such as isopropanol or ethylene glycol may serve in sanitizing and cosmetic formulations, they are incompatible with food and drinks.

Finally, discontinuing the production of both natural and synthetic ethanol for fuel blends would represent a setback in the transition to renewable energy sources, as fossil fuels are not sustainable in the long term.

An Opinion on Ethanol Consumption

The widespread popularity of ethanol consumption is perplexing, given its well-documented harmful effects. While misinformation may be a cause, it is also plausible that alcohol consumption among young people often begins as an expression of rebellion or as a means to cope with emotions, largely due to its disinhibitory effects and the initial feelings of well-being it induces (Drabwell et al., 2020).

However, not all individuals consume alcohol responsibly and in moderation. On the contrary, alcohol abuse is prevalent, and its short- and long-term consequences for the consumer are significant. For example, acute intoxication can result from an overdose in the short term, while recurrent consumption may cause serious health conditions such as cirrhosis or kidney failure over time (Medical News Today, 2024).

Alcohol abuse is associated with social stigmas and negative societal consequences, as it is frequently linked to crimes, felonies, and domestic violence (Graham & Livingston, 2011). Additionally, alcohol abuse is a major contributing factor to fatal road accidents (Barry et al., 2022; Love et al., 2023). Severe alcoholism not only endangers the consumer due to injuries sustained in traffic accidents but also increases the likelihood of becoming an indirect victim of road incidents or crimes committed by intoxicated individuals.

Furthermore, a significant concern related to ethanol consumption is the presence of methanol, a highly toxic alcohol. Methanol is not produced through the fermentation of sugars by yeasts but rather in small quantities through the fermentation of pectins in the must. Generally, most alcoholic beverages contain trace amounts of methanol that the body can tolerate. However, alcoholic beverages produced from red fruits, such as grapes, cherries, and plums, contain considerably higher levels of methanol. For example, apricot brandy contains up to 10.81 mg methanol per 1 ml of ethanol, plum brandy up to 8.85 mg methanol per 1 ml of ethanol, and another variety of plum brandy up to 5.29 mg methanol per 1 ml of ethanol. In contrast, corn and barley distillates, such as Scotch whisky, contain only 0.080–0.260 mg methanol per 1 ml of ethanol (World Health Organization, 1988). Therefore, moderating the consumption of such alcoholic beverages is advisable to minimize methanol intake.

Additionally, consumers must ensure that their alcoholic beverages are obtained from legitimate sources, as the real danger lies in adulterated products distributed on the black market. These counterfeit beverages often contain dangerously high levels of methanol and other toxic alcohols, such as ethylene glycol, which are added to mimic the taste of authentic drinks. Even moderate consumption of these substances can result in severe consequences, including blindness or death (Magnúsdóttir et al., 2010).

An Opinion on the Use of Ethanol as Fuel

Ethanol has a dark side despite its potential benefits as fuel. Bioethanol fuel has significant environmental drawbacks, making it a less eco-friendly option. The monoculture farming required for bioethanol production usually demands substantial water, fertilizer, and pesticide inputs.

The water footprint of bioethanol production is significant. For instance, it is estimated that producing one ton of bioethanol requires 209 m^3 of water from sugarcane crops, 133 m^3 from sugar beet crops, and 1,222 m^3 from maize crops (Gerbens-Leenes & Hoekstra, 2012). Consequently, bioethanol production is highly dependent on water availability and land allocation for cultivation. This dependency poses a challenge, as drinking water reserves decrease annually (Salehi, 2022), while the global population continues to grow and human activities exacerbate aquifer contamination.

Similarly, the production of synthetic ethanol from natural gas presents environmental issues comparable to those of the petroleum industry, primarily concerning harm to living organisms, air, water, soil, and subsoil (Assaf et al., 2024).

Therefore, further efforts are required to reduce the environmental impact of ethanol production. Moreover, it is necessary to improve engine performance to make ethanol competitive with standard fuels like gasoline and diesel.

Final Comments on Ethanol

The high demand for alcoholic beverages remains ethanol's most evident use. However, ethanol also serves various purposes in the food industry. As the only food-grade alcohol approved for consumption, it is used as an eluent or additive in concentrates, flavorings, and syrups (European Commission, 2010).

Ethanol is essential as a sanitizing and cleaning agent; it is commonly found in first aid kits in the form of denatured alcohol. Its compatibility with human skin also makes it suitable for use in cosmetic products such as perfumes and deodorants. Furthermore, ethanol is a recognized green solvent for synthesis.

Ethanol is also a molecule of interest in renewable energy, particularly for its potential as a biofuel. The demand for bioethanol alternative fuels is expected to increase in the coming years due to the growing challenges associated with natural gas extraction and oil refining.

In summary, ethanol has long been valued for its taste and effects when consumed. In medicine, it serves as an effective sanitizer, while in the energy sector, it is gaining recognition as a fuel for motor vehicles. These diverse applications make ethanol one of the most versatile compounds. As society strives for a greener future, ethanol is evolving from a recreational beverage into a sustainable fuel source.

Some Curiosities About Ethanol

- *Saccharomyces cerevisiae* strains are used to ferment sugars to produce alcohol and are also used as baker's yeast (Legras et al., 2007).
- Ethanol molecules drift among the stars; one possible origin is the condensation of carbon dioxide and ethane that are irradiated by stars at -262.15°C (Schriver et al., 2007).
- Since 1855, non-drinkable ethanol has been formulated to prevent poisoning. Denatured ethanol consists of adding small amounts of compounds that initially only introduce an unpleasant smell and taste and cannot be separated by distillation. Over the years, denaturing agents such as camphor, turpentine, acetic acid, methanol, pyridine, acetone, methylene blue, aniline blue, naphthalene, castor oil, and benzene, among others, have been used. Paradoxically, some of these compounds are more harmful to health than ethanol itself ("Denatured Alcohol," 1931).
- Snake Venom is the world's strongest beer in 2024, with an ABV of 67.5%. Brewmeister has been producing this strong ale-style beverage in Scotland since October 2013 (Riddell, 2013).

- Throughout history, high-alcohol spirits have been crafted worldwide. Three current offerings include Spirytus Rektyfikowany vodka (96% ABV), produced by Polmos in Poland, as well as Everclear (95.6% ABV) and Golden Grain (95% ABV), both produced by Luxco in St. Louis, Missouri, USA (Freedman, 2023).
- In a study investigating alcohol and sugar addiction in mice, researchers discovered that female mice are more prone than male mice to developing an addiction to these substances (Wei et al., 2021).
- Alcohol consumption causes approximately 2.6 million deaths worldwide, accounting for 4.7% of all deaths globally (World Health Organization, 2025).
- In the postmortem analysis of 175 fatal cases of ethanol intoxication, the mean blood alcohol concentration was found to be 355 mg/100 ml (Heatley & Crane, 1990). In the USA, the legal limit for blood alcohol concentration is 80 mg/100 ml for drivers over 21 years old (Fell & Voas, 2014).
- According to a study conducted in the USA during the COVID-19 pandemic, 60% of people surveyed increased their consumption of alcoholic beverages, while only 12% decreased it compared to their pre-pandemic consumption. Among those who increased consumption, the main reasons were stress, availability of drinks, and boredom (Grossman et al., 2020).
- Since 1860, the German engineer and inventor Nicolaus August Otto (1832–1891) contemplated the design of engines that could run on ethanol (Songstad et al., 2009).
- In 1979, the Brazilian Fiat 147 was the first modern commercial automobile capable of running solely on pure ethanol fuel (E100) (Stellantis, 2025).
- In the Holy Bible, 21st Century King James Version, the word "wine" appears 325 times. The first instance is in the Book of Genesis (9:21): "And he drank of the wine, and was drunk: and he was uncovered within his tent". Also, in John (2:1-11), it is mentioned that Jesus changed water into wine (*Holy Bible. The 21st Century King James Version*, 1994).
- Dionysus, in Greek mythology—known as Bacchus in Roman mythology—was one of the twelve Olympian gods. He was a deity associated with wine, as well as fertility, pleasure, nature, theater, festivity, and ritual madness (Mac Góráin, 2017).

- In Mexica mythology, the *Centzon Totochtin*, or "Four Hundred Rabbits," were a group of 400 spirits, divine rabbits, or gods of drunkenness. They are associated with sleep and awakening, obfuscation and lucidity, as well as death (López Austin, 1994).

References

Abdulrazeq, H. F., Ali, R., Najib, H., Doberstein, C., Oyelese, A., Gokaslan, Z., Malik, A. N., Asaad, W. F., & Greenblatt, S. (2024). Al-Zahrawi (936–1013 AD): On the surgical treatment of neurological disorders by the father of operative surgery. *World Neurosurgery*, *184*, 236-240.e1. https://doi.org/10.1016/j.wneu.2024.01.169.

Adger, H., & Saha, S. (2013). Alcohol use disorders in adolescents. *Pediatrics in Review*, *34*(3), 103–114. https://doi.org/10.1542/pir.34-3-103.

Álvarez-Ríos, G. D., Casas, A., Pérez-Volkow, L., Figueredo-Urbina, C. J., de Dios Páramo-Gómez, J., & Vallejo, M. (2022). Pulque and pulquerías of Mexico City: A traditional fermented beverage and spaces of biocultural conservation. *Journal of Ethnic Foods*, *9*(1), 40. https://doi.org/10.1186/s42779-022-00155-2.

Assaf, J. C., Mortada, Z., Rezzoug, S.-A., Maache-Rezzoug, Z., Debs, E., & Louka, N. (2024). Comparative review on the production and purification of bioethanol from biomass: A focus on corn. *Processes*, *12*(5), 1001. https://doi.org/10.3390/pr12051001.

Baraona, E., Abittan, C. S., Dohmen, K., Moretti, M., Pozzato, G., Chayes, Z. W., Schaefer, C., & Lieber, C. S. (2001). Gender differences in pharmacokinetics of alcohol. *Alcoholism: Clinical and Experimental Research*, *25*(4), 502–507. https://doi.org/10.1111/j.1530-0277.2001.tb02242.x.

Barry, V., Schumacher, A., & Sauber-Schatz, E. (2022). Alcohol-impaired driving among adults—USA, 2014–2018. *Injury Prevention*, *28*(3), 211–217. https://doi.org/10.1136/injuryprev-2021-044382.

Berg, A. T., Dixon-Salazar, T., Meskis, M. A., Danese, S. R., Le, N. M. D., & Perry, M. S. (2024). Caregiver-reported outcomes with real-world use of cannabidiol in Lennox-Gastaut syndrome and Dravet syndrome from the BECOME survey. *Epilepsy Research*, *200*, 107280. https://doi.org/10.1016/j.eplepsyres.2023.107280.

Britannica. (2024, November 16). *Prohibition United States history [1920–1933]*. Https://Www.Britannica.Com/Event/Prohibition-United-States-History-1920-1933..

Britton, R. A. (1969). *Direct hydration of ethylene to ethanol* (Patent US3686334A).

Capello, C., Fischer, U., & Hungerbühler, K. (2007). What is a green solvent? A comprehensive framework for the environmental assessment of solvents. *Green Chemistry*, *9*(9), 927. https://doi.org/10.1039/b617536h.

Carravetta, V., Gomes, A. H. de A., Marinho, R. dos R. T., Öhrwall, G., Ågren, H., Björneholm, O., & de Brito, A. N. (2022). An atomistic explanation of the ethanol–water azeotrope. *Physical Chemistry Chemical Physics*, *24*(42), 26037–26045. https://doi.org/10.1039/D2CP03145K.

Castellano Actual. (2015, January 7). *Duda resuelta: Alcohol*. Https://Www.Udep.Edu.Pe/Castellanoactual/Duda-Resuelta-Alcohol/.

Charters, S. (2006). The history of wine. In S. Charters (Ed.), *Wine and Society* (pp. 10–45). Elsevier. https://doi.org/10.1016/B978-0-7506-6635-0.50006-9.

Chen, C. C., Liaw, H. J., Shu, C. M., & Hsieh, Y. C. (2010). Autoignition temperature data for methanol, ethanol, propanol, 2-butanol, 1-butanol, and 2-methyl-2,4-pentanediol. *Journal of Chemical & Engineering Data*, *55*(11), 5059–5064. https://doi.org/10.1021/je100619p.

Chrzanowski, G. (2020). Saccharomyces cerevisiae—An interesting producer of bioactive plant polyphenolic metabolites. *International Journal of Molecular Sciences*, *21*(19), 7343. https://doi.org/10.3390/ijms21197343.

Ciani, M., Comitini, F., & Mannazzu, I. (2013). Fermentation. In S. E. Jørgensen & B. D. Fath (Eds.), *Encyclopedia of Ecology* (Vol. 2, pp. 1548–1557). Elsevier. https://doi.org/10.1016/B978-008045405-4.00272-X.

Crosland, M. P. (1978). *Gay-Lussac*. Cambridge University Press. https://doi.org/10.1017/CBO9780511564390.

del Campo, M., & Jones, K. L. (2017). A review of the physical features of the fetal alcohol spectrum disorders. *European Journal of Medical Genetics*, *60*(1), 55–64. https://doi.org/10.1016/j.ejmg.2016.10.004.

Denatured alcohol. (1931). *Industrial & Engineering Chemistry*, *23*(1), 2–3. https://doi.org/10.1021/ie50253a005.

Drabwell, L., Eng, J., Stevenson, F., King, M., Osborn, D., & Pitman, A. (2020). Perceptions of the use of alcohol and drugs after sudden bereavement by unnatural causes: Analysis of online qualitative data. *International Journal of Environmental Research and Public Health*, *17*(3), 677. https://doi.org/10.3390/ijerph17030677.

Edenberg, H. J. (2007). The genetics of alcohol metabolism: Role of alcohol dehydrogenase and aldehyde dehydrogenase variants. *Alcohol Res Health*, *1*(5), 5–13.

Escalante, A., López Soto, D. R., Velázquez Gutiérrez, J. E., Giles-Gómez, M., Bolívar, F., & López-Munguía, A. (2016). Pulque, a traditional Mexican alcoholic fermented beverage: Historical, microbiological, and technical aspects. *Frontiers in Microbiology*, *7*, 1–18. https://doi.org/10.3389/fmicb.2016.01026.

European Commission. (2010). *Directive 2009/32/EC of the European Parliament and of the Council of 23 April 2009 on the Approximation of the Laws of the Member States on Extraction Solvents Used in the Production of Foodstuffs and Food Ingredients*.

Fell, J. C., & Voas, R. B. (2014). The effectiveness of a 0.05 blood alcohol concentration (BAC) limit for driving in the United States. *Addiction*, *109*(6), 869–874. https://doi.org/10.1111/add.12365.

Freedman, B. (2023, September 19). *Where to Find 12 of the World's Strongest Liquors*. Https://Www.Foodandwine.Com/Drinks/12-Worlds-Strongest-Liquors.

Geeks for Geeks. (2024, November 16). *List of Countries Where Alcohol is Banned*. Https://Www.Geeksforgeeks.Org/Countries-Where-Alcohol-Is-Banned/.

Gerbens-Leenes, W., & Hoekstra, A. Y. (2012). The water footprint of sweeteners and bio-ethanol. *Environment International*, *40*, 202–211. https://doi.org/10.1016/j.envint.2011.06.006.

Graham, K., & Livingston, M. (2011). The relationship between alcohol and violence: Population, contextual and individual research approaches. *Drug and Alcohol Review*, *30*(5), 453–457. https://doi.org/10.1111/j.1465-3362.2011.00340.x.

Grossman, E. R., Benjamin-Neelon, S. E., & Sonnenschein, S. (2020). Alcohol consumption during the COVID-19 pandemic: A cross-sectional survey of US adults. *International Journal of Environmental Research and Public Health*, *17*(24), 9189. https://doi.org/10.3390/ijerph17249189.

Hahn-Hägerdal, B., Galbe, M., Gorwa-Grauslund, M. F., Lidén, G., & Zacchi, G. (2006). Bio-ethanol – the fuel of tomorrow from the residues of today. *Trends in Biotechnology*, *24*(12), 549–556. https://doi.org/10.1016/j.tibtech.2006.10.004.

Heatley, M. K., & Crane, J. (1990). The blood alcohol concentration at post-mortem in 175 fatal cases of alcohol intoxication. *Medicine, Science and the Law*, *30*(2), 101–105. https://doi.org/10.1177/002580249003000203.

Hennell, H. (1828). XIX. On the mutual action of sulphuric acid and alcohol, and on the nature of the process by which ether is formed. *Philosophical Transactions of the Royal Society of London*, *118*, 365–371. https://doi.org/10.1098/rstl.1828.0021.

Holy Bible. The 21st Century King James Version. (1994). Deuel Enterprises.

Ingledew, W. M., & Hysert, D. W. (1994). Brewing Technology. In *Reference Module in Food Science* (pp. 1–11). Elsevier. https://doi.org/10.1016/B978-0-08-100596-5.02926-7.

Insider Monkey. (2024, November 16). *5 Countries That Produce The Most Alcohol*. Https://Www.Insidermonkey.Com/Blog/5-Countries-That-Produce-the-Most-Alcohol-1081658/.

Institute of Alcohol Studies. (2020, October). *The Alcohol Industry: an Overviewing*. Https://Www.Ias.Org.Uk/Wp-Content/Uploads/2020/12/The-Alcohol-Industry-%E2%80%93-An-Overview.Pdf.

Jaywant, S. A., Singh, H., & Arif, K. M. (2022). Sensors and instruments for brix measurement: A review. *Sensors*, *22*(6), 2290. https://doi.org/10.3390/s22062290.

Jensen, W. B. (2004). The origin of alcohol proof. *Journal of Chemical Education*, *81*(9), 1258. https://doi.org/10.1021/ed081p1258.

Jönsson, P. G. (1976). Hydrogen bond studies. CXIII. The crystal structure of ethanol at 87 K. *Acta Crystallographica Section B Structural Crystallography and Crystal Chemistry*, *32*(1), 232–235. https://doi.org/10.1107/S0567740876002653.

Jorge, M., Gomes, J. R. B., & Barrera, M. C. (2022). The dipole moment of alcohols in the liquid phase and in solution. *Journal of Molecular Liquids*, *356*, 119033. https://doi.org/10.1016/j.molliq.2022.119033.

Kessler, F. H., von Diemen, L., Ornell, F., & Sordi, A. O. (2021). Cannabidiol and mental health: possibilities, uncertainties, and controversies for addiction treatment. *Brazilian Journal of Psychiatry*, *43*(5), 455–457. https://doi.org/10.1590/1516-4446-2021-1838.

Kizasu, S., Sato, T., Inoue, Y., Tasaki, H., Shirasuna, K., Okiishi, Y., & Iwata, H. (2024). Effect of low ethanol concentration in maturation medium on developmental ability, mitochondria, and gene expression profile in mouse oocytes. *Reproductive Biology*, *24*(2), 100854. https://doi.org/10.1016/j.repbio.2023.100854.

Legras, J.-L., Merdinoglu, D., Cornuet, J.-M., & Karst, F. (2007). Bread, beer and wine: *Saccharomyces cerevisiae* diversity reflects human history. *Molecular Ecology*, *16*(10), 2091–2102. https://doi.org/10.1111/j.1365-294X.2007.03266.x.

Liu, L., Zhang, Y., Ge, W., Lin, Z., Sinnott-Armstrong, N., & Yang, L. (2024). Revealing the 2300-year-old fermented beverage in a bronze bottle from shaanxi, China. *Fermentation*, *10*(7), 365. https://doi.org/10.3390/fermentation10070365.

López Austin, A. (1994). *El conejo en la cara de la luna* (A. López Austin, Ed.; 1st ed.). Era.

Love, S., Rowland, B., & Davey, J. (2023). Exactly how dangerous is drink driving? An examination of vehicle crash data to identify the comparative risks of alcohol-related crashes. *Crime Prevention and Community Safety*, *25*(2), 131–147. https://doi.org/10.1057/s41300-023-00172-6.

Luo, X., Kranzler, H. R., Zuo, L., Lappalainen, J., Yang, B., & Gelernter, J. (2006). ADH4 gene variation is associated with alcohol dependence and drug dependence in European Americans: Results from HWD tests and case–control association studies. *Neuropsychopharmacology*, *31*(5), 1085–1095. https://doi.org/10.1038/sj.npp.1300925.

Mac Góráin, F. (2017, September 27). *Dionysus*. Oxford Bibliographies Online Datasets. https://doi.org/10.1093/obo/9780195389661-0284.

Magnúsdóttir, K., Kristinsson, J., & Jóhannesson, Þ. (2010). Lækningar og saga: Svikið áfengi. *Læknablaðið*, *2010*(10), 626–628. https://doi.org/10.17992/lbl.2010.10.322.

Mainkar, P., Ray, A., & Chandrasekhar, S. (2024). Solvents: From past to present. *ACS Omega*, *9*(7), 7271–7276. https://doi.org/10.1021/acsomega.3c07508.

Marisa Mendes. (2017). *Distillation - Innovative Applications and Modeling* (M. F. Mendes, Ed.; 1st ed.). InTech. https://doi.org/10.5772/62970.

McCaul, M. E., Roach, D., Hasin, D. S., Weisner, C., Chang, G., & Sinha, R. (2019). Alcohol and women: A Brief overview. *Alcoholism: Clinical and Experimental Research*, *43*(5), 774–779. https://doi.org/10.1111/acer.13985.

Medical News Today. (2024, November 16). *What effects does alcohol have on health?* Https://Www.Medicalnewstoday.Com/Articles/305062.

Merk. (2024, November 16). *Ethanol*. Https://Www.Merckmillipore.Com/MX/En/Product/Ethanol,MDA_CHEM-100983.

Moure-Rodríguez, L., & Caamano-Isorna, F. (2020). We need to delay the age of onset of alcohol consumption. *International Journal of Environmental Research and Public Health*, *17*(8), 2739. https://doi.org/10.3390/ijerph17082739.

Murachman, B., Pranantyo, D., & Sandjaya Putra, E. (2014). Study of gasohol as alternative fuel for gasoline substitution: Characteristics and performances. *International Journal of Renewable Energy Development*, *3*(3), 175–183. https://doi.org/10.14710/ijred.3.3.175-183.

Narro, G. E. C., Díaz, L. A., Ortega, E. K., Garín, M. F. B., Reyes, E. C., Delfin, P. S. M., Arab, J. P., & Bataller, R. (2024). Alcohol-related liver disease: A global perspective. *Annals of Hepatology*, *29*(5), 101499. https://doi.org/10.1016/j.aohep.2024.101499.

Nelson, D. J., Brammer, C., & Li, R. (2009). Substituent effects in acid-catalyzed hydration of alkenes, measured under consistent reaction conditions. *Tetrahedron Letters*, *50*(47), 6454–6456. https://doi.org/10.1016/j.tetlet.2009.08.128.

OPEC. (2024, November 16). *Counting the cost of speculation.* Www.Opec.Org/ Opec_web/En/Press_room/2907.Htm.

Ornoy, A., & Ergaz, Z. (2010). Alcohol abuse in pregnant women: Effects on the fetus and newborn, mode of action and maternal treatment. *International Journal of Environmental Research and Public Health*, *7*(2), 364–379. https://doi.org/ 10.3390/ijerph7020364.

Pérez-García, J. M., Suárez-Suárez, S., Doallo, S., & Cadaveira, F. (2022). Effects of binge drinking during adolescence and emerging adulthood on the brain: A systematic review of neuroimaging studies. *Neuroscience & Biobehavioral Reviews*, *137*, 104637. https://doi.org/10.1016/j.neubiorev.2022.104637.

Pietraszek, A., Gregersen, S., & Hermansen, K. (2010). Alcohol and type 2 diabetes. A review. *Nutrition, Metabolism and Cardiovascular Diseases*, *20*(5), 366–375. https://doi.org/10.1016/j.numecd.2010.05.001.

Pouliot, S., Liao, K. A., & Babcock, B. A. (2018). Estimating willingness to pay for E85 in the United States using an intercept survey of flex motorists. *American Journal of Agricultural Economics*, *100*(5), 1486–1509. https://doi.org/10.1093/ajae/aay041.

Rasmussen, S. C. (2019). From aqua vitae to E85: The history of ethanol as fuel. *Substantia*, *3*(2), 43–55.

Reiman, A. (2009). Cannabis as a substitute for alcohol and other drugs. *Harm Reduction Journal*, *6*(1), 35. https://doi.org/10.1186/1477-7517-6-35.

Riddell, R. (2013, October 24). *Strongest beer in the world? Snake Venom sports 67.5% abv.* Https://Www.Fooddive.Com/News/Strongest-Beer-in-the-World-Snake-Venom-Sports-675-Abv/185830/.

Ritchie, H., & Roser, M. (2024, January). *Alcohol Consumption.* Https:// Ourworldindata.Org/Alcohol-Consumption.

Salehi, M. (2022). Global water shortage and potable water safety; today's concern and tomorrow's crisis. *Environment International*, *158*, 106936. https://doi.org/10.1016/ j.envint.2021.106936.

Salguero, A., Marengo, L., Cendán, C. M., Morón, I., Ruiz-Leyva, L., & Pautassi, R. M. (2024). Ethanol drinking at adulthood is sensitive to S1-R antagonism and is promoted by binge ethanol self-administration at adolescence. *Drug and Alcohol Dependence*, *260*, 111338. https://doi.org/10.1016/j.drugalcdep.2024.111338.

Schmidt-Rohr, K. (2015). Why combustions are always exothermic, yielding about 418 kJ per mole of O_2. *Journal of Chemical Education*, *92*(12), 2094–2099. https://doi.org/ 10.1021/acs.jchemed.5b00333.

Schriver, A., Schriver-Mazzuoli, L., Ehrenfreund, P., & d'Hendecourt, L. (2007). One possible origin of ethanol in interstellar medium: Photochemistry of mixed CO2–C2H6 films at 11K. A FTIR study. *Chemical Physics*, *334*(1–3), 128–137. https://doi.org/10.1016/j.chemphys.2007.02.018.

Scinska, A., Koros, E., Habrat, B., Kukwa, A., Kostowski, W., & Bienkowski, P. (2000). Bitter and sweet components of ethanol taste in humans. *Drug and Alcohol Dependence*, *60*(2), 199–206. https://doi.org/10.1016/S0376-8716(99)00149-0.

Sewell, S. L. (2014). The spatial diffusion of beer from its sumerian origins to today. In M. Patterson & N. Hoalst-Pullen (Eds.), *The Geography of Beer* (pp. 23–29). Springer Netherlands. https://doi.org/10.1007/978-94-007-7787-3_3.

Silvello, G. C., Bortoletto, A. M., de Castro, M. C., & Alcarde, A. R. (2021). New approach for barrel-aged distillates classification based on maturation level and machine learning: A study of cachaça. *LWT, 140*, 110836. https://doi.org/10.1016/j.lwt.2020.110836.

Songstad, D. D., Lakshmanan, P., Chen, J., Gibbons, W., Hughes, S., & Nelson, R. (2009). Historical perspective of biofuels: learning from the past to rediscover the future. *In Vitro Cellular & Developmental Biology - Plant, 45*(3), 189–192. https://doi.org/10.1007/s11627-009-9218-6.

Sorbi, D., McGill, D. B., Thistle, J. L., Therneau, T. M., Henry, J., & Lindor, K. D. (2000). An assessment of the role of liver biopsies in asymptomatic patients with chronic liver test abnormalities. *The American Journal of Gastroenterology, 95*(11), 3206–3210. https://doi.org/10.1111/j.1572-0241.2000.03293.x.

Spectral Database for Organic Compounds. (2024, November 16). *SDBS Information.* Https://Sdbs.Db.Aist.Go.Jp/.

Statista. (2024, November 16). *Ranking de marcas de bebidas alcohólicas con más valor a nivel mundial en 2024.* Https://Es.Statista.Com/Estadisticas/618005/Marcas-de-Bebidas-Espirituosas-Lideres-Del-Mundo-Por-Valor-de-Marca/).

Stellantis. (2025, January 12). *Fiat celebra los 45 años del primer coche propulsado por etanol del mundo.* Https://Www.Media.Stellantis.Com/Es-Es/Fiat/Press/Fiat-Celebra-Los-45-Anos-Del-Primer-Coche-Propulsado-Por-Etanol-Del-Mundo.

Stikkers, D. E. (2002). Octane and the environment. *Science of The Total Environment, 299*(1–3), 37–56. https://doi.org/10.1016/S0048-9697(02)00271-1.

Suarez-Bertoa, R., Zardini, A. A., Keuken, H., & Astorga, C. (2015). Impact of ethanol containing gasoline blends on emissions from a flex-fuel vehicle tested over the Worldwide harmonized Light duty Test Cycle (WLTC). *Fuel, 143*, 173–182. https://doi.org/10.1016/j.fuel.2014.10.076.

Subramaniyan, V., Chakravarthi, S., Jegasothy, R., Seng, W. Y., Fuloria, N. K., Fuloria, S., Hazarika, I., & Das, A. (2021). Alcohol-associated liver disease: A review on its pathophysiology, diagnosis and drug therapy. *Toxicology Reports, 8*, 376–385. https://doi.org/10.1016/j.toxrep.2021.02.010.

Synthetic ethanol capacity on rise. (1966). *Chemical & Engineering News Archive, 44*(30), 30. https://doi.org/10.1021/cen-v044n030.p030.

Thomasson, H. R. (2002). Gender differences in alcohol metabolism. In M. Galanter, H. Begleiter, R. Deitrich, D. Gallant, D. Goodwin, E. Gottheil, A. Paredes, M. Rothschild, D. Thiel, & H. Edwards (Eds.), *Recent Developments in Alcoholism* (pp. 163–179). Springer US. https://doi.org/10.1007/0-306-47138-8_9.

Union Carbide Corporation. (2024, November 16). *History.* Https://Www.Unioncarbide.Com/History.Html.

Wade, L. G. (2017). *Organic Chemistry* (9th ed.). Pearson.

Warren, S., & Wyatt, P. (2004). *Organic Synthesis: The Disconnection Approach* (2nd ed.). Wiley.

Wei, S., Hertle, S., Spanagel, R., & Bilbao, A. (2021). Female mice are more prone to develop an addictive-like phenotype for sugar consumption. *Scientific Reports, 11*(1), 7364. https://doi.org/10.1038/s41598-021-86797-9.

Williams, D. B. G., & Lawton, M. (2010). Drying of organic solvents: Quantitative evaluation of the efficiency of several desiccants. *The Journal of Organic Chemistry*, *75*(24), 8351–8354. https://doi.org/10.1021/jo101589h.

Wilson, D. F., & Matschinsky, F. M. (2020). Ethanol metabolism: The good, the bad, and the ugly. *Medical Hypotheses*, *140*, 109638. https://doi.org/10.1016/j.mehy.2020.109638.

Wisniak, J. (2009). Eugène Melchior Peligot. *Educación Química*, *20*(1), 61–69. https://doi.org/10.1016/S0187-893X(18)30008-9.

World Health Organization. (1988). IARC Monographs on the Evaluation of Carcinogenic Risks to Humans. In *International Agency for Research on Cancer* (Vol. 44, p. 425). World Health Organization, Distribution and Sales Service.

World Health Organization. (2004). *Global Status Report: Alcohol Policy*.

World Health Organization. (2025, January 11). *Over 3 million annual deaths due to alcohol and drug use, majority among men*. Https://Www.Who.Int/News/Item/25-06-2024-over-3-Million-Annual-Deaths-Due-to-Alcohol-and-Drug-Use-Majority-among-Men.

Zaso, M. J., Goodhines, P. A., Wall, T. L., & Park, A. (2019). Meta-analysis on associations of alcohol metabolism genes with alcohol dse Disorder in East Asians. *Alcohol and Alcoholism*, *54*(3), 216–224. https://doi.org/10.1093/alcalc/agz011.

Zehra, A., Burns, J., Liu, C. K., Manza, P., Wiers, C. E., Volkow, N. D., & Wang, G.-J. (2018). Cannabis addiction and the brain: A review. *Journal of Neuroimmune Pharmacology*, *13*(4), 438–452. https://doi.org/10.1007/s11481-018-9782-9.

Zhang, M., & Yu, Y. (2013). Dehydration of ethanol to ethylene. *Industrial & Engineering Chemistry Research*, *52*(28), 9505–9514. https://doi.org/10.1021/ie401157c.

Zheng, X. W., & Han, B. Z. (2016). Baijiu (白酒), Chinese liquor: History, classification and manufacture. *Journal of Ethnic Foods*, *3*(1), 19–25. https://doi.org/10.1016/j.jef.2016.03.001.

Chapter 3

Aspirin

"When a regular person gets sick, they take an aspirin. When a writer gets sick, they take notes."
Chuck Palahniuk

The History of Aspirin (*Acetylsalicylic acid*)

Aspirin is nothing other than a molecule that results from observation, meticulous analysis, and the persistent dedication of scientists whose discoveries spanned centuries. The journey of aspirin's development was extensive, shaped by contributions that far exceeded the capabilities of a single lifetime. Instead, this synthetic compound represents the cumulative knowledge inherited and expanded upon by several generations of scientists, many of whom remain unnamed. This chapter explores the background and applications of this renowned medication, initially known as acetylsalicylic acid and later renamed aspirin.

The analgesic properties of willow bark have been recognized since ancient Egypt, approximately 3,500 years ago (Montinari et al., 2019). Beyond the Egyptians, records indicate that other ancient civilizations, such as the

Greeks and Assyrians, also utilized willow (*Salix spp.*) bark infusions for their anti-inflammatory and analgesic effects.

The modern history of acetylsalicylic acid synthesis is marked by discrepancies and certain ambiguities, particularly regarding the dates of key events in the 19th century. Due to the lack of consensus, consulting additional references to compare the sequence of discoveries is advisable. However, irrespective of exact dates, the contributions of the scientists discussed below remain widely recognized.

One of the earliest contributions to aspirin's history in the modern era came from Edward Stone (1702–1768), an English clergyman and natural philosopher. In 1763, Stone presented the first clinical documentation of willow bark's medicinal properties, describing its antipyretic effects in cases of ague (Pierpoint, 1997).

Decades later, between 1824 and 1826, Italian scientists Francesco Fontana (1794–1867) and Bartolomeo Rigatelli (Marson & Pasero, 2011) advanced the identification of salicin, a glucoside of *ortho*-hydroxybenzyl alcohol (chemical formula $C_{13}H_{18}O_7$) and the precursor to salicylic acid, the active compound in willow bark. Subsequently, between 1826 and 1827, German pharmacist Johann Andreas Buchner (1783–1852) successfully isolated small quantities of pure salicin crystals. Shortly thereafter, in 1829, Henri Leroux (1797–1871) refined salicin to a higher purity and confirmed its antipyretic properties (Wood, 2015). Recent studies on purified willow bark extracts, particularly from white willow (*Salix alba*) and black willow (*Salix nigra*), highlight salicin's potential therapeutic benefits, including palliative effects for patients with rheumatoid arthritis (Lin et al., 2023).

In 1838, Raffaele Piria (1814–1865), a young Italian physician and chemist, devised a method to produce salicylic acid from salicin. He demonstrated that salicin decomposes into D-glucose and saligenin, which can then be oxidized to yield salicylic acid. The salt sodium salicylate had been used for fever treatment toward 1875. However, its side effects prompted the search for derivatives with improved therapeutic properties and reduced toxicity (Gąsowska-Bajger et al., 2023).

The synthesis of acetylsalicylic acid was first achieved in 1852 by the distinguished French scientist Charles Gerhardt (1816–1856). He synthesized the compound by condensing salicylic acid with acetic anhydride, discovering that the acetylated derivative was less toxic than salicylic acid and exhibited promising pharmacological properties (Lévesque & Lafont, 2000).

The history of acetylsalicylic acid continues in Germany, where chemists Adolph Wilhelm Hermann Kolbe (1818–1884) and Rudolf Schmitt (1830–

1898) collaborated on the total synthesis of salicylic acid around 1859. Both scientists were interested and encouraged by the pharmacological benefits of acetylsalicylic acid, so they dedicated part of their research to synthesizing the precursor salicylic acid. Finally, they developed a method to produce salicylic acid from phenol without using salicin as a precursor, i.e., via a synthetic route. Years later, in 1885, Schmitt patented the synthesis system. This synthesis was relevant for acetylsalicylic acid production, so it was named the Kolbe-Schmitt reaction (Fernández Braña et al., 2005).

In 1897, the German pharmacist Felix Hoffmann (1868–1946), working under the guidance of chemist Arthur Eichengrün (1867–1949) at Bayer, optimized the synthesis of acetylsalicylic acid. Building upon the foundational work of Gerhardt, Kolbe, and Schmitt, Hoffmann's efforts were driven by Bayer's commercial aims. The company's decisive actions positioned acetylsalicylic acid as a cornerstone of 20th-century medicine. Two key events for the industrial production of aspirin were the registration of Aspirin® in Germany in 1899 and the US patent in 1900 that facilitated its commercialization first as a powder and subsequently in tablet form (Bayer de México, 2024). Bayer's strategic promotion of aspirin as the leading choice for analgesic and antipyretic therapies significantly expanded its global reach.

Bayer is also credited with coining the term "aspirin." In Germany, salicylic acid was known as *spirsäure* (spiric acid in English). A name inspired by the meadowsweet plant, formerly known as *Spiraea ulmaria* (now *Filipendula ulmaria*), from which salicin and salicylates were obtained. The name aspirin was therefore derived from this botanical connection (Sokolova et al., 2024). Over time, the widespread use of this medication led to acetylsalicylic acid and aspirin becoming interchangeable terms in both scientific and everyday language.

Throughout the 20th century, aspirin remained a prolific subject of scientific inquiry, particularly in applied science. In 1982, pharmacologist Sir John Vane (1927–2004) was awarded the Nobel Prize in Physiology or Medicine for elucidating the anti-inflammatory mechanisms of acetylsalicylic acid. His research uncovered how aspirin inhibits the production of prostaglandins and other chemical mediators involved in the inflammatory response, such as prostacyclin, a lipid molecule that promotes vasodilation and inhibits smooth muscle growth and platelet aggregation (Harding, 2004).

Today, aspirin is marketed in numerous pharmaceutical formulations, including Aspirin®, Zorprin®, Aspro®, and ASA®. Additionally, compounded medications combine aspirin with other active ingredients. For example, Cafiaspirin®, Anacin®, and Aspircaf® pair aspirin with caffeine, while Supac®, Vanquish®, and Excedrin® combine acetaminophen, aspirin, and caffeine. Alka-Seltzer® includes aspirin, citric acid, and sodium bicarbonate.

General Information on Aspirin

Acetylsalicylic acid is a stable white solid. It has the chemical formula $C_9H_8O_4$, a molecular weight of 180.15 g/mol, a melting point of 136°C, and a density of 1.40 g/mL. Acetylsalicylic acid is slightly soluble in water, with a solubility of approximately 3 mg/mL at 20°C. However, the solubility in water increases significantly when the pH exceeds the pKa value of 5.3 due to the formation of its corresponding salt. For example, the reaction of salicylic acid with NaOH forms sodium acetylsalicylate ($C_9H_7NaO_4$), which has a molecular weight of 202.14 g/mol and a melting point of 208.5–213.5°C. This organic salt is significantly more soluble in water (approximately 1000 g/L). For this reason, sodium acetylsalicylate is a common ingredient in pharmaceutical formulations. Aspirin or acetylsalicylic acid is also known as 2-ethanoatebenzoic acid, 2-acetoxybenzoic acid, and ortho-acetylsalicylic acid.

The Safety Data Sheet (SDS) for acetylsalicylic acid (CAS 50-78-2) indicates that it is a stable compound but should be stored in a dry place. If inhaled, move to an area with fresh air (Sigma-Aldrich, 2024). In case of contact with the skin or eyes, rinse immediately with water. If accidentally ingested, drink water and monitor for the following symptoms: tinnitus, nausea, vomiting, drowsiness, confusion, and rapid breathing. In such cases, medical professionals should provide assistance.

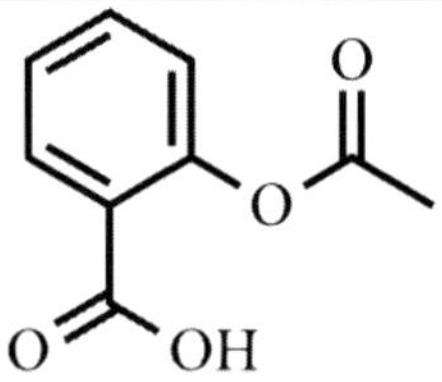	**ASPIRIN**	
	Other names: Acetylsalicylic acid, 2-ethanoatebenzoic acid, 2-acetoxybenzoic acid, and ortho-acetylsalicylic acid	
	Origin: Synthetic	Formula: $C_9H_8O_4$
Type: Salicylate	Discoverer: NA	MW: 180.15 g/mol
Appearance: White solid	Inventor: Charles Gerhardt in 1852	mp: 136°C
Uses: Antipyretic, anti-inflammatory, and antiplatelet	Toxicity: Medicament	bp: NA

Treatment comprises the oral administration of activated charcoal and the intravenous administration of fluids and bicarbonate (O'Malley & O'Malley, 2022). For more information, see the section 'Pharmaceutical applications and pharmacology of aspirin.'

The Structure and Reactivity of Salicylates and Aspirin

Aspirin belongs to the salicylate group because it is a derivative of salicylic acid (Furman, 2018), a molecule with a six-membered aromatic ring containing a carboxyl (-COOH) and a hydroxyl (-OH) group. These functional groups exhibit different chemical reactivity because the carbon atom in the -COOH group acts as an electrophilic center, whereas the oxygen atom in the -OH group is nucleophilic. This difference enables the selective synthesis of derivatives.

Other compounds of interest derived from salicylic acid are, for example, methyl salicylate, ethyl salicylate, salsalate, tri-salicylic acid, diflunisal, and mesalazine. These compounds exhibit diverse chemical properties. For example, methyl salicylate (Api et al., 2024) and ethyl salicylate (Lapczynski et al., 2007) are ingredients in fragrance formulations. Tri-salicylic acid (MedChemExpress, 2024), salsalate (Trifilio et al., 2021), diflunisal (Lawton & Chapman, 1993), and mesalazine (Cipolla et al., 2002) share anti-inflammatory properties with salicylic acid.

Despite being synthesized more than a century ago, the structure of aspirin remains an active area of research. Recent studies have investigated the crystallization methods of acetylsalicylate salts and their characterization using single-crystal X-ray diffraction (Búdová et al., 2018; Kolev et al., 2009). The crystal structure of acetylsalicylic acid, determined using the SXD neutron time-of-flight Laue diffractometer, exhibits monoclinic symmetry and belongs to the space group $P2_1/c$ (Harrison et al., 2003).

The spectroscopic characterization of acetylsalicylic acid unequivocally confirms its connectivity and structure (AIST, 2024). ^{1}H-NMR ($CDCl_3$) shows four aromatic proton signals at 8.12, 7.62, 7.36, and 7.14 ppm; one signal for the carboxylic acid proton at 11.77 ppm; and one signal for the acetyl proton at 2.35 ppm. ^{13}C-NMR ($CDCl_3$) shows six signals corresponding to the aromatic ring at 151.28, 134.90, 132.51, 126.17, 124.01, and 122.26 ppm; one signal for the carboxylic acid carbon at 170.20 ppm; and two signals for the acetate group at 169.76 and 20.99 ppm.

In the infrared (IR) spectrum (KBr disc) of aspirin, the bands of carboxylic acid are strong, with the O–H stretching vibration at 3006 cm^{-1} and the C=O stretching vibration at 1693 cm^{-1}. Additionally, the intense C=O stretching vibration of the acetate appears at 1754 cm^{-1}.

The reactivity of acetylsalicylic acid and its salt derivatives is well known; the most significant reaction is deacetylation, which regenerates salicylic acid. However, the reactivity of aspirin is more closely related to its pharmacological activity, specifically its metabolic pathways upon ingestion. For example, acetylsalicylic acid can undergo hydrolysis and oxidation to form gentisic acid through a mechanism that involves the hydroxyl radical (•OH) and cytochrome P450 (Cyt-P450). Gentisic acid in the body forms derivatives associated with anti-inflammatory activity, such as methyl gentisate, salicytamide, and gentisamide (Borges & Castle, 2015). The section 'Pharmaceutical applications and pharmacology of aspirin' discusses additional plausible mechanisms of aspirin's action.

The Synthesis of Aspirin

Currently, the synthesis of acetylsalicylic acid follows a method similar to the classical Kolbe-Schmitt reaction (Hessel et al., 2006) and begins with the deprotonation of phenol. Under conditions of 125°C and 100 atm, sodium phenoxide reacts with CO_2, leading to carboxylation of the aromatic ring at the *ortho* position. Treating sodium salicylate with an acidic solution yields neutral salicylic acid.

Then, the salicylic acid undergoes acetylation with acetic anhydride or acetyl chloride, producing acetylsalicylic acid as the neutral condensation product. When acetylsalicylic acid reacts with an alkaline sodium hydroxide solution, it forms sodium acetylsalicylate, a water-soluble salt suitable for pharmaceutical formulations (Figure 6).

The total synthesis pathway is currently the primary method for obtaining acetylsalicylic acid, as the molecule is small, and reagents are readily available. Additionally, each reaction during the process has a high yield, which is more environmentally sustainable in comparison to the synthesis of other drugs.

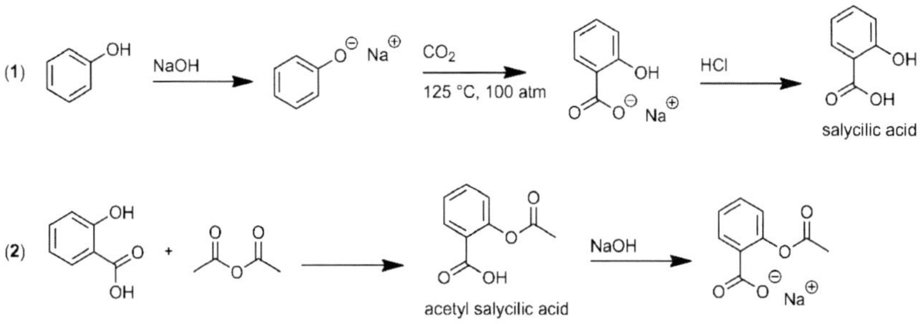

Figure 6. Schematic representation of the synthesis pathway of aspirin from phenol.

The Pharmaceutical Applications and Pharmacology of Aspirin

The classification of aspirin as a prodrug is well-documented. Drug regulatory sources offer valuable insights into its role as an active ingredient, given its numerous applications (FDA, 2024a). Aspirin possesses antipyretic, non-narcotic analgesic, non-steroidal anti-inflammatory, and anticoagulant properties. This section outlines its mechanisms of action, contraindications, and side effects.

Aspirin irreversibly inhibits the activity of cyclooxygenase enzymes 1 and 2 (COX-1 and COX-2), thereby reducing the production of prostaglandin and thromboxane precursors. At the peripheral level, aspirin inhibits prostaglandin synthesis, preventing the stimulation of bradykinin pain receptors. This mechanism provides an analgesic and anti-inflammatory effect, while as an antiplatelet agent, aspirin inhibits the enzyme COX-1. This decrease in thromboxane synthesis prevents platelet aggregation and activation, thus reducing platelet deposition in the bloodstream.

Aspirin is indicated for the treatment of various conditions, including pain (e.g., headache, dental pain, menstrual cramps, muscle pain, and low back pain), fever, non-rheumatic inflammation (e.g., musculoskeletal pain, bursitis, capsulitis, tendonitis, acute tenosynovitis, and Kawasaki disease), arthritis (e.g., rheumatoid arthritis, juvenile arthritis, and osteoarthritis), and thrombosis (e.g., thrombophlebitis, arterial thrombosis, and postoperative thromboembolism).

Aspirin is administered orally, preferably after meals, while ingestion before breakfast should be avoided. Most aspirin treatments are recommended for individuals over 16 years of age and contraindicated in children under 12. Additionally, aspirin is contraindicated in cases of severe liver failure, severe kidney failure, and pregnancy. Additionally, aspirin should be avoided during breastfeeding, as it is excreted in breast milk (Vademecum, 2024a).

Patients receiving high or moderate doses of aspirin for prolonged periods may exhibit elevated serum alanine aminotransferase (ALT) levels. This elevation may increase bilirubin levels and indicate liver injury, in which case, the administration of aspirin should be discontinued. Long-term aspirin use may cause symptoms such as nausea, anorexia, abdominal pain, and encephalopathy (due to increased bilirubin), as well as signs of liver dysfunction, including hyperammonemia and coagulopathy.

Adverse reactions associated with aspirin include an increased risk of bleeding (perioperative bleeding, gastrointestinal bleeding, or hematuria), vomiting, hematomas, epistaxis, hypoprothrombinemia, rhinitis, paroxysmal bronchospasm, severe dyspnea, asthma, nasal congestion, dyspepsia, gastric or duodenal ulcer, urticaria, rash, angioedema, and pruritus (Asociación Española de Pediatría, 2024).

When combined with other drugs, aspirin may alter their effects by interfering with, enhancing, or inhibiting them. For instance, it enhances the potency and toxicity of acetazolamide, increases nephrotoxicity when combined with cyclosporine, and raises the risk of ototoxicity with vancomycin. It also heightens the risk of renal failure when used with diuretics and increases plasma concentrations of barbiturates, digoxin, phenytoin, lithium salts, zidovudine, valproic acid, and methotrexate. Furthermore, aspirin enhances the effect of insulin and sulfonylureas, while decreasing the efficacy of interferon alpha, beta-blocker antihypertensives, and uricosurics. Metamizole may also reduce aspirin's antiplatelet effect.

Although aspirin is available over the counter, its numerous contraindications and potential drug interactions require physician supervision. It should only be administered as part of a comprehensive treatment plan and avoid self-medication.

The Aspirin Industry

The Bayer brand continues manufacturing pharmaceutical products containing acetylsalicylic acid as an active ingredient. For instance, Alka-Seltzer plus® was the best-selling medicine in the United States for treating colds, allergies, and sinuses, with 30.5 million units sold in 2016 (Statista, 2024). Global aspirin production reached 41,000 tons in 2022, generating revenue of USD

2,356.55 million (ChemAnalyst, 2024). Sales are expected to grow by 2.70%, with profits of USD 2,917.59 million by 2030 (Zion Market Research, 2024).

Major players in the aspirin market include Allegiant Health (USA), Bal Pharma (India), JQC (Huayin) Pharmaceutical (China), Cardinal Health (USA), Par Pharmaceutical (USA), Industria Química Andina y Cia (Colombia), Perrigo Company (USA), Nanjing Pharmaceutical Factory (China), Trumac Healthcare (India), Globela Pharma (India), and LNK International (USA). Currently, Allegiant Health manufactures drugs for Bayer using aspirin as an active ingredient (Allegiant Health, 2024).

Possible Consequences of Discontinuing Aspirin

If aspirin production were to cease, it would have significant economic and medical consequences.

From an economic perspective, given aspirin's widespread global use, its removal from the market would likely drive up the prices of alternative medications. The cost of certain aspirin substitutes—prescribed as analgesics, antipyretics, or anticoagulants—would increase, particularly for medications still under patent protection. Like other industries, the pharmaceutical sector incurs substantial expenses for research and production, costs which are included in the final price. Thus, the economic impact of an aspirin ban would be considerable.

In a world without aspirin, alternative medications would be necessary. While aspirin substitutes offer certain advantages, they also have drawbacks. On the one hand, newer medications may exhibit greater specificity and enhanced pharmacological potency, but on the other hand, their long-term side effects may still be unknown. In contrast, aspirin has been extensively studied from chemical, biochemical, pharmacological, and clinical perspectives, and its side effects are well-documented.

Fortunately, aspirin-like medications remain available, including over-the-counter options and others with expired patents (FDA, 2024b; Vademecum, 2024b). For severe muscle pain, non-steroidal, non-psychotropic analgesics such as ibuprofen or paracetamol may be prescribed. Additionally, alternatives to Alka-Seltzer® for heartburn include sodium bicarbonate, calcium carbonate, aluminum hydroxide, magnesium hydroxide, bismuth subsalicylate, and simethicone (polydimethylsiloxane). For antiplatelet therapy, triflusal and ticlopidine can serve as substitutes. Consequently, alternative medications effectively fulfill all of aspirin's primary functions.

Although aspirin may no longer be considered an innovative medication, it was indispensable in the early 20th century, when fewer pharmaceutical options were available and much of the world's population lacked access to modern medicine. At that time, possessing a pack of aspirin could mean the difference between life and death.

The Significance of Aspirin in Medicine and Drug Synthesis

In retrospect, the isolation of the precursor salicin, extracted from willow and other plants, demonstrated that neither the whole plant nor the complete extract was necessary to obtain a product with medicinal effects. Today, we know that the isolation of active ingredients enables the design of effective medications, while non-therapeutic substances are removed. Moreover, the purification of active components prolongs the medication's shelf life. This approach has allowed researchers to focus on improving the properties of future bioactive compounds, such as bioactivity, solubility, stability, and selectivity (Atanasov et al., 2021).

Particularly, aspirin represents a triumph of modern science in both organic synthesis and medicine. The synthesis of aspirin validated the work of chemists who first isolated its precursors from natural extracts, modified salicylic acid to produce acetylsalicylic acid, and synthesized the drug without reliance on natural sources. This reliable multi-therapeutic medication provided unprecedented benefits for physicians and patients in the early 20th century.

As a model drug, aspirin helped researchers understand the importance of synthesizing pharmaceutical derivatives from naturally derived molecules. In the clinic, aspirin has eased the treatment of several diseases and health conditions. Therefore, as long as ailments such as fever or muscle pain remain part of life, carrying a bottle of aspirin on hand will be a relief.

Some Curiosities About Aspirin

- Acetylation of alcohols and phenols with either acetic anhydride or acetyl chloride has been one of the most important derivatization reactions for obtaining compounds with enhanced pharmacological activity. In addition to aspirin (derived from salicylic acid), other examples include ethyl acetate (from ethanol), acetophenone (from benzene), paracetamol (from 4-aminophenol), and heroin (from morphine), among other examples.

- During World War I, the Germans cut off supplies of aspirin to the UK; nevertheless, the British found a supplier in 1917: the Australian pharmacist George Nicholas, who named his drug Aspro and founded a pharmaceutical company (Walker, 2025). Years later, Bayer purchased the Aspro brand.
- At the end of World War I, the Allies took control of the aspirin patent as part of the war reparations (Moya, 2015).
- After World War I, Bayer tried to regain the rights to aspirin but was unsuccessful. Beginning in 1925 and continuing through World War II, Bayer, AGFA, BASF, and other German companies were part of the IG Farben conglomerate, which was linked to the Nazi regime. This association damaged the reputation of these companies for years (López-Muñoz et al., 2009).
- During World War II, both sides supplied aspirin to their soldiers to relieve the pain of wounds and reduce fever caused by infections (Mann & Plummer, 1991).
- Bayer aspirin was the first medicine on the Moon. It was included in the first aid kit that NASA prepared for the crew of the Apollo 11 mission, which successfully landed Neil Armstrong, the first man on the Moon, on July 20, 1969 (Gargantilla, 2023).
- Since the first edition in 1977, aspirin has been included on the World Health Organization's Essential Drug List (World Health Organization, 2025).
- In addition to its regular use as a medicine, long-term, low-dose aspirin administration is associated with a slight to moderate reduction in the risk of certain types of cancer (Skriver et al., 2024).
- For many decades, aspirin was the flagship product of the pharmaceutical company Bayer. In fact, the 1991 *Guinness Book of World Records* recognized it as the best-selling painkiller in the world (Bayer de México, 2024).
- Some studies indicate that taking aspirin after coronary bypass surgery increases the chance of recovery and reduces the risk of postoperative complications (Mangano, 2002).

References

AIST. (2024, November 25). *Spectral Database for Organic Compounds SDBS.* Https://Sdbs.Db.Aist.Go.Jp.

Allegiant Health. (2024, November 25). *HealthA2Z.* Https://Allegiant-Health.Com/ Products/Healtha2z/.

Api, A. M., Bartlett, A., Belsito, D., Botelho, D., Bruze, M., Bryant-Freidrich, A., Burton, G. A., Cancellieri, M. A., Chon, H., Dagli, M. L., Dekant, W., Deodhar, C., Farrell, K., Fryer, A. D., Jones, L., Joshi, K., Lapczynski, A., Lavelle, M., Lee, I., … Tokura, Y. (2024). RIFM fragrance ingredient safety assessment, methyl salicylate, CAS registry number 119-36-8. *Food and Chemical Toxicology, 194,* 115047. https://doi.org/10.1016/j.fct.2024.115047.

Asociación Española de Pediatría. (2024, November 25). *Ácido acetilsalicílico (AAS).* Https://Www.Aeped.Es/Comite-Medicamentos/Pediamecum/Acido-Acetilsalicilico-Aas.

Atanasov, A. G., Zotchev, S. B., Dirsch, V. M., & Supuran, C. T. (2021). Natural products in drug discovery: advances and opportunities. *Nature Reviews Drug Discovery, 20*(3), 200–216. https://doi.org/10.1038/s41573-020-00114-z.

Bayer de México. (2024, November 25). *Aspirina.* Https://Www.Aspirina.Com.Mx/.

Borges, R. S., & Castle, S. L. (2015). The antioxidant properties of salicylate derivatives: A possible new mechanism of anti-inflammatory activity. *Bioorganic & Medicinal Chemistry Letters, 25*(21), 4808–4811. https://doi.org/10.1016/j.bmcl.2015.07.001.

Búdová, M., Skořepová, E., & Čejka, J. (2018). Sodium aspirin salts: Crystallization and characterization. *Crystal Growth & Design, 18*(9), 5287–5294. https://doi.org/ 10.1021/acs.cgd.8b00718.

ChemAnalyst. (2024, March). *Decode the Future of Aspirin.* Https://Www. Chemanalyst.Com/Industry-Report/Aspirin-Market-3149.

Cipolla, G., Crema, F., Sacco, S., Moro, E., De Ponti, F., & Frigo, G. (2002). Nonsteroidal anti-inflammatory drugs and inflammatory bowel disease: Current perspectives. *Pharmacological Research, 46*(1), 1–6. https://doi.org/10.1016/S1043-6618(02)00033-6.

FDA. (2024a, November 25). *Aspirin: Questions and Answers.* Https://Www.Fda. Gov/Drugs/Safe-Use-Aspirin/Aspirin-Questions-and-Answers.

FDA. (2024b, November 25). *Drug Safety and Availability.* Https://Www.Fda. Gov/Drugs/Drug-Safety-and-Availability.

Fernández Braña, M., Río Álvarez, L. A. del, Trives Lombardero, C., & Salazar Sánchez, N. (2005). La verdadera historia de la Aspirina. *Real Academia Nacional de Farmacia, 71*(19), 813–819.

Furman, B. L. (2018). Salicylic Acid. In *Reference Module in Biomedical Sciences* (pp. 1–5). Elsevier. https://doi.org/10.1016/B978-0-12-801238-3.97758-4.

Gargantilla, P. (2023, August 24). *El primer medicamento que acompañó a los astronautas a la Luna.* Https://Www.Muyinteresante.Com/Salud/61255.Html.

Gąsowska-Bajger, B., Sosnowska, K., Gąsowska-Bodnar, A., & Bodnar, L. (2023). The effect of acetylsalicylic acid, as a representative non-steroidal anti-inflammatory drug, on the activity of myeloperoxidase. *Pharmaceuticals*, *16*(7), 1012. https://doi.org/10.3390/ph16071012.

Harding, A. (2004). Sir John Robert Vane. *The Lancet*, *364*(9451), 2090. https://doi.org/10.1016/S0140-6736(04)17571-5.

Harrison, A., Ibberson, R., Robb, G., Whittaker, G., Wilson, C., & Youngson, D. (2003). In situ neutron diffraction studies of single crystals and powders during microwave irradiation. *Faraday Discuss.*, *122*, 363–379. https://doi.org/10.1039/B203379H.

Hessel, V., Löb, P., & Löwe, H. (2006). Industrial and real-life applications of micro-reactor process engineering for fine and functional chemistry. In H.-K. Rhee, I.-S. Nam, & J. M. Park (Eds.), *Studies in Surface Science and Catalysis* (Vol. 159, pp. 35–46). Elsevier. https://doi.org/10.1016/S0167-2991(06)81535-1.

Kolev, T., Koleva, B. B., Seidel, R. W., Spiteller, M., & Sheldrick, W. S. (2009). Benzamidinium acetylsalicylate: Crystal structure of the first salt with acetylsalicylate anion. *Structural Chemistry*, *20*(3), 533–536. https://doi.org/10.1007/s11224-009-9423-2.

Lapczynski, A., McGinty, D., Jones, L., Bhatia, S., Letizia, C. S., & Api, A. M. (2007). Fragrance material review on ethyl salicylate. *Food and Chemical Toxicology*, *45*(1), S397–S401. https://doi.org/10.1016/j.fct.2007.09.043.

Lawton, G. M., & Chapman, P. J. (1993). Diflunisal — a long-acting non-steroidal anti-inflammatory drug. *Australian Dental Journal*, *38*(4), 265–271. https://doi.org/10.1111/j.1834-7819.1993.tb05494.x.

Lévesque, H., & Lafont, O. (2000). L'aspirine à travers les siècles: Rappel historique. *La Revue de Médecine Interne*, *21*, S8–S17. https://doi.org/10.1016/S0248-8663(00)88720-2.

Lin, C.-R., Tsai, S. H. L., Wang, C., Lee, C.-L., Hung, S.-W., Ting, Y.-T., & Hung, Y. C. (2023). Willow bark (Salix spp.) used for painrelief in arthritis: A meta-analysis of randomized controlled trials. *Life*, *13*(10), 2058. https://doi.org/10.3390/life13102058.

López-Muñoz, F., García-García, P., & Alamo, C. (2009). The pharmaceutical industry and the German National Socialist Regime: I. G. Farben and pharmacological research. *Journal of Clinical Pharmacy and Therapeutics*, *34*(1), 67–77. https://doi.org/10.1111/j.1365-2710.2008.00972.x.

Mangano, D. T. (2002). Aspirin and mortality from coronary bypass surgery. *New England Journal of Medicine*, *347*(17), 1309–1317. https://doi.org/10.1056/NEJMoa020798

Mann, C. C., & Plummer, M. L. (1991). *The Aspirin Wars: Money, Medicine, and l00 Years of Rampant Competition*. Knopf.

Marson, P., & Pasero, G. (2011). The Italian contributions to the history of salicylates. *Reumatismo*, *58*(1). https://doi.org/10.4081/reumatismo.2006.66.

MedChemExpress. (2024, November 25). *Tri-Salicylic acid* . Https://Www.Medchemexpress.Com/Tri-Salicylic-Acid.Html?Locale=es-ES.

Montinari, M. R., Minelli, S., & De Caterina, R. (2019). The first 3500 years of aspirin history from its roots – A concise summary. *Vascular Pharmacology*, *113*, 1–8. https://doi.org/10.1016/j.vph.2018.10.008.

Moya, A. S. (2015, July 2). *La aspirina, de botín de los aliados tras la Primera Guerra Mundial a "propiedad de toda la humanidad."* Https://Www.Abc.Es/Salud/Noticias/20150522/Abci-Origen-Aspirina-Medicamento-201505141952.Html.

O'Malley, G. F., & O'Malley, R. (2022, September). *Intoxicación por aspirina (ácido acetilsalicílico).* Https://Www.Msdmanuals.Com/Es/Hogar/Traumatismos-y-Envenenamientos/Intoxicaciones-o-Envenenamientos/Intoxicaci%C3%B3n-Por-Aspirina-%C3%A1cido-Acetilsalic%C3%ADlico?Ruleredirectid=757.

Pierpoint, W. S. (1997). Edward Stone (1702–1768) and Edmund Stone (1700–1768): Confused identities resolved. *Notes and Records of the Royal Society of London, 51*(2), 211–217. https://doi.org/10.1098/rsnr.1997.0018.

Sigma-Aldrich. (2024, July 9). *Safety Data Sheet, Acetylsalicylic acid.* Https://Www. Sigmaaldrich.Com/MX/En/Sds/Sigma/A5376?UserType=undefined.

Skriver, C., Maltesen, T., Dehlendorff, C., Skovlund, C. W., Schmidt, M., Sørensen, H. T., & Friis, S. (2024). Long-term aspirin use and cancer risk: a 20-year cohort study. *JNCI: Journal of the National Cancer Institute, 116*(4), 530–538. https://doi.org/ 10.1093/jnci/djad231.

Sokolova, E., Krol, T., Adamov, G., Minyazeva, Y., Baleev, D., & Sidelnikov, N. (2024). Total content and composition of phenolic compounds from Filipendula genus plants and their potential health-promoting properties. *Molecules, 29*(9), 2013. https://doi.org/10.3390/molecules29092013.

Statista. (2024, November 25). *Unit sales of the leading cold, allergy and sinus tablet/packet brands in the United States in 2016.* Https://Www. Statista.Com/Statistics/462788/Us-Unit-Sales-of-the-Leading-Cold-Tablet-Brands/.

Trifilio, S., Gordon, L., Rubin, H., Grosshans, N., & Mehta, J. (2021). The non-steroidal anti-inflammatory drug salsalate provides safe and effective control of mucositis-unrelated pain during autologous and allogeneic hematopoietic stem cell transplantation. *Supportive Care in Cancer, 29*(7), 3643–3648. https://doi.org/ 10.1007/s00520-020-05664-x.

Vademecum. (2024a, November 25). *Aspirina 500 mg Comprimidos.* Https://Www. Vademecum.Es/Espana/Medicamento/298/Aspirina-500-Mg-Comprimidos.

Vademecum. (2024b, November 25). *Vademecum.* Https://Www.Vademecum.Es/.

Walker, R. (2025, January 15). *Nicholas, George Richard Rich (1884 - 1960).* Https://Www.Eoas.Info/Biogs/P002464b.Htm.

Wood, J. N. (2015). From plant extract to molecular panacea: A commentary on Stone (1763) 'An account of the success of the bark of the willow in the cure of the agues.' *Philosophical Transactions of the Royal Society B: Biological Sciences, 370*(1666), 20140317. https://doi.org/10.1098/rstb.2014.0317.

World Health Organization. (2025, January 8). *Model List of Essential Medicines.* Https://List.Essentialmeds.Org/.

Zion Market Research. (2024, November 25). *Aspirin Market Size, Share, Trends, Growth 2030.* Https://Www.Zionmarketresearch.Com/Report/Aspirin-Market.

Chapter 4

Ammonia

"With a dab of ammonia
She once revived me when I fainted one day!"
Toshie Nohara

The History of Ammonia

Ammonia is continuously synthesized in nature as a byproduct of organic matter decomposition in all living organisms, from microalgae and bacteria to large mammals. In this sense, ammonia is an essential part of the nitrogen cycle, a process in which inert nitrogen moves from the atmosphere and land to living organisms through various metabolic processes before being reincorporated into the air or mineral substrates (Francis et al., 2007). Nitrogen gas is a very abundant molecule in the atmosphere (78%), but most living beings cannot use it. Hence, the fixation of nitrogen into ammonia and nitrates depends on microorganisms such as bacteria (Dellagi et al., 2020), fungi (Koch et al., 2021), and archaea (Mowafy et al., 2022). During the nitrogen cycle, ammonia and ammonium salts are oxidized to nitrites and nitrates that plants use to form part of their biomass in the form of amino acids, nucleic acids, and proteins. The nitrogen cycle continues when plants are consumed, allowing organisms to incorporate organic nitrogen into their structures. At the end of their life cycle, these organisms release organic nitrogen, which can once again be metabolized by microorganisms. Part of the organic matter undergoes denitrification, releasing nitrogen gas, while another portion regenerates ammonia and ammonium salts through ammonification, thereby continuing the nitrogen cycle.

The name of the molecule ammonia became widely recognized in Europe by the late 18th century, although alchemists had long been familiar with it (Wentrup, 2024). The name derives from a mineral called "ammoniak" (German for ammonia), which contains ammonium chloride (Harper, 2025). The mineral's name originates from deposits near the Temple of Jupiter Ammon in Libya, where it was first extracted. In antiquity, Greek settlers established colonies near the temple in North Africa around 630 BC. Thus, the etymology of ammonia reflects influences from ancient Egyptian and Greek civilizations. In Greek, "ammōniakón" (ἀμμωνιακόν) means "belonging to Ammon," an ancient Egyptian deity often depicted with ram's horns. The Romans later adopted this god as a syncretic deity, identifying him with Jupiter Ammon (Encyclopedia Mythica, 2025). The name Ammon is believed to derive from the Greek word "ammos" (ἄμμος), meaning "sand."

Pliny the Elder (20–79 AD) and Lucius Junius Moderatus Columella (4–70 AD) mentioned "Hammoniacus sal," which likely referred to smelling salts or baking salts containing ammonium chloride or ammonium carbonate. These salts were used in ceremonies, medicinal preparations, stimulants, perfume ingredients, and even in the manufacture of porcelain (McCrory, 2006).

Historical records suggest that ammonia derivatives were used in Asia during the Tang Dynasty (618–907 AD), particularly in metallurgy and medicine. Ammonia salts were a product traded along the Silk Road (Sutton et al., 2020). The exploitation and use of ammonia derivatives in solid or liquid forms (particularly ammonium salts) was evident in ancient times; however, the methods for isolating and identifying the elements comprising ammonia gas remained unknown. The elemental composition of ammonia was discovered during the first studies in mineralogy and chemistry.

One of the first scientists to study ammonia was Joseph Priestley (1733–1804), who called the gas "alkaline air." He first shared his findings in a private letter to Benjamin Franklin in 1773 and later to the Royal Society (Sutton et al., 2020). Another key figure, French chemist Claude Louis Berthollet (1748–1822), made a breakthrough in 1785 when he identified ammonia gas as consisting of three hydrogen atoms and one nitrogen atom (Lemay & Oesper, 1946).

Throughout the 18th and 19th centuries, scientists successfully synthesized ammonia gas, though only in small quantities and mainly for research purposes, such as the synthesis of amines and other organic precursors that laid the foundation for modern organic chemistry.

Almost as important as the synthesis of ammonia gas itself were the advances made by the German scientists Fritz Haber (1868–1934) and Carl Bosch (1874–1940), who completed their research on the synthesis and large-scale production of ammonia from hydrogen and nitrogen gases in a relatively short time. While Haber and Bosch played leading roles, other German scientists, including Walther Nernst (1864–1941), Gabriel van Oordt, and Friedrich Jost, also contributed to early research on ammonia synthesis.

Haber's work may have been supported by earlier findings from researchers such as Nernst and Jost, who investigated the equilibria and synthesis of ammonia from hydrogen and nitrogen under extreme conditions—685°C, 50 atm, and with various catalysts such as platinum, iron, and manganese. Despite Nernst and Jost efforts, these experiments achieved only very low yields (Johnson, 2022a). Over the years, Haber conducted numerous investigations into gas mixtures, building extensive knowledge on the subject that ultimately allowed him to refine his process. Around 1904, Haber, in collaboration with van Oordt (Sheppard, 2017), attempted the synthesis of ammonia from nitrogen and hydrogen, testing various catalysts under high-temperature conditions. However, without applying high pressure, they achieved yields of less than 1% (Haber & van Oordt, 1905). Haber may have considered abandoning the project at one point, but by modifying the reaction conditions—particularly by increasing the system's pressure—he achieved promising results. Encouraged by these improvements, Haber persevered in improving the process (Berl, 1937).

The German chemical company Badische Anilin-& Soda-Fabrik (BASF) played a key role in industrializing ammonia production by bridging the gap between Haber's academic research and Bosch's engineering expertise. The formation of this team is in part due to the vision of Carl Engler (1842–1925), a member of BASF's supervisory board, who recognized the potential of Haber's work and Bosch's involvement. Once the process of producing ammonia was economically viable, BASF facilitated patent filings and supported the transition to industrial-scale ammonia production.

While Bosch led the engineering process of ammonia, he also supervised the process for hydrogen gas from water vapor and coal, which presented a double challenge and, in this case, a double success.

The collaborator Alwin Mittasch (1869–1953) discovered an efficient iron-based catalyst that improved ammonia conversion rates. And Franz Lappe (1878–1950) focused on the technical aspects of the process. Years later, Lappe served as the head of the Mechanical Engineering Department at the Oppau plant from 1925 to 1944, where he contributed to the industrial implementation of these advancements.

Several patents were filed during the development of the Haber-Bosch process, documenting the various stages of its evolution:

- Bosch applied for the first patent, “Process of Producing Ammonia” (US41884808A), on March 2, 1908. This patent described the production of ammonia from titanium cyanonitride, obtained by heating a mixture of titanic acid and carbon in the presence of nitrogen (Bosch, 1908).
- Haber and his collaborator Robert le Rossignol (1888–1976) filed the patent application “Production of Ammonia” (US1202995A) on August 13, 1909, describing the use of a catalyst to synthesize ammonia from hydrogen and nitrogen (Haber & Le Rossignol, 1909).
- Bosch, Mittasch, Hans Wolf, and Georg Stern filed the patent “Catalytic Agent for Use in the Production of Ammonia” (US1148570A) on December 24, 1910, which specified the use of an iron catalyst in the ammonia synthesis process (Bosch et al., 1910).
- Bosch and Mittasch filed the patent “Production of Ammonia” (US1158167A) on November 18, 1911, which outlined optimized reaction conditions, including an iron oxide catalyst (Bosch & Mittasch, 1911).

By 1913, BASF constructed the first ammonia production plant in Oppau, Germany, with an initial production capacity of approximately 11,000 tons per year. This novel process rendered traditional methods of synthesizing and extracting ammonium compounds obsolete. The culmination of this industrial project transformed the world we know today, which is dependent on synthetic ammonia. Initially, the technological advancements enabled by the Haber-Bosch process allowed Germany to become self-sufficient in producing synthetic ammonia-based salts, supporting agriculture and the production of explosives during World War I (Johnson, 2022b). However, this was only the first time that synthetic ammonia took part in historical events—and its influence continues to this day.

The scientific community recognized the groundbreaking nature of Haber’s work, awarding him the Nobel Prize in Chemistry in 1918 for his synthesis of ammonia from its elements. The following year, Walther Nernst was awarded the Nobel Prize in Chemistry for his contributions to thermochemistry, which included work on ammonia synthesis. Finally, in 1931, Carl Bosch shared the Nobel Prize in Chemistry with Friedrich Bergius (1884–1949) for their pioneering work on high-pressure chemical methods, securing their place in industrial chemistry history.

The Synthesis of Ammonia

Approximately 70% of the total ammonia gas production is derived from natural gas through two industrial processes at high temperatures and pressures. The first step involves steam methane reforming (Figure 7.1.), where natural gas, a mixture of light hydrocarbons primarily composed of methane, is used to produce hydrogen gas. In the second step, the Haber-Bosch process (Figure 7.2) combines this hydrogen gas with nitrogen gas in an exothermic reaction to form ammonia gas, with a reaction enthalpy of $\Delta H = -46.1$ kJ/mol (Wang et al., 2024). Usually, the process is carried out at temperatures ranging from 200 to 400 atm and temperatures ranging from 400 to 600°C. Technical details on the engineering principles behind this method can be found in the literature (Berwal et al., 2021).

Modern research aims to optimize the Haber-Bosch process by reducing the time and increasing conversion rates, thereby minimizing energy costs. Despite its industrial significance, the ammonia process typically achieves conversion rates of less than 15% (Humphreys et al., 2021). Thus, there is a wide margin to improve the conversion rates, particularly developing new advanced catalysts; for example, using metal oxide catalysts such as iron oxides (Fe_2O_3, Fe_3O_4, or $Fe_{1-x}O$), aluminum oxide (Al_2O_3), barium oxide (BaO), and potassium oxide (K_2O). While any innovation tries to increase efficiency and yield, the fundamental principles of the Haber-Bosch process—relying on high pressure and temperature—remain unchanged.

Currently, due to the growing demand for methane to meet the needs of ammonia production and domestic fuel consumption (Valera-Medina et al., 2021), the production of syngas (synthetic methane) has increased through a process known as coal steam reforming or coke steam reforming (Yüzbaşıoğlu et al., 2021). This process involves using coal as a source of H_2, which is a more economical option due to the abundance of coal reserves and the lower extraction costs compared to natural gas.

$$(1)\quad \text{Natural gas} + \text{Air} + H_2O(l) \xrightarrow[\text{"Steam methane reforming"}]{\text{Electricity}} H_2(g) + CO_2(g)$$

$$(2)\quad 3H_2(g) + N_2(g) \xrightarrow[\text{"Haber-Bosch process"}]{\text{200-400 atm, 400-600 °C}} 2NH_3(g)$$

$$(3)\quad NH_4Cl(ac) + NaOH(ac) \longrightarrow NH_3(ac) + NaCl(ac) + H_2O(l)$$

$$(4)\quad NH_3(ac) \xrightarrow[\text{CaO trap}]{\Delta T} NH_3(g)$$

$$(5)\quad N_2(g) + 6Li(s) \longrightarrow 2Li_3N(s)$$

$$(6)\quad Li_3N(s) + 3H_2(g) \longrightarrow NH_3(g) + 3LiH(s)$$

Figure 7. Reaction schemes for hydrogen production through steam methane reforming (1), ammonia synthesis using the Haber-Bosch process (2), and alternative laboratory ammonia synthesis methods (3–6).

Alternatives for obtaining ammonia in the laboratory usually start with solid salt reagents, such as ammonium chloride or ammonium sulfate since handling solids and liquids is safer. Aqueous ammonia is easily obtained via an acid-base reaction (Martin, 1999), for example, by mixing ammonium chloride and sodium hydroxide in aqueous medium (Figure 7.3). When the resulting aqueous ammonia (formally in equilibrium with ammonium hydroxide) is heated, ammonia gas is released. The drying of ammonia gas can then be achieved using a calcium oxide trap, after which the gas is collected for further use (Figure 7.4). Another method for producing ammonia in the laboratory involves the formation of lithium nitride (Li_3N) by reacting a lithium-based alloy with nitrogen gas (Yamaguchi et al., 2020) (Figure 7.5). When Li_3N reacts with H_2, it yields ammonia (Figure 7.6). While these laboratory-scale methods are effective for small-scale applications, they are irrelevant at the industrial level because the Haber-Bosch process is profitable and efficient for large-scale production.

The Physical and Chemical Properties of Ammonia

At room temperature, ammonia is a colorless gas with a strong, penetrating, nauseating odor. It is also named anhydrous ammonia, nitrogen trihydride, ammonia gas, azane, and spirit of hartshorn.

Ammonia has a melting point of −77.7°C and a boiling point of −33.3°C. At 20°C, the vapor pressure of ammonia gas is 8.5 atm. The critical temperature of ammonia is 132°C. Ammonia is soluble in water (89.9 g/100 mL at 0°C) and self-dissociates into ammonium ions (NH_4^+) and amide ions (NH_2^-); however, its equilibrium constant is small ($K = 10^{-30}$). In solution, ammonia has a pKa of 9.24, classifying it as a weak base. The density of ammonia gas is 0.73 kg/m^3, and its relative density compared to air is 0.6.

According to the hazardous materials standard NFPA 704, ammonia (CAS 7664-41-7) has the following ratings: health = 0 (blue), flammability = 1 (red), instability = 0 (yellow), and no special hazards indicated (white). Therefore, the main hazard is health-related (Cameo Chemicals, 2025). Its high toxicity is attributed to its high solubility in water and its ability to disperse into an aqueous or water-rich medium, such as the human body (which is approximately 70% water). Studies indicate that the release of aqueous ammonia into the sea leads to high concentrations of ammonium hydroxide, a toxic substance harmful to aquatic life (Lin et al., 2022; Zhang et al., 2023).

The effects of ammonia on the human body and recommended actions in case of exposure are detailed below. Ammonia is toxic and corrosive, causing damage to the skin, eyes, respiratory tract, and nasal mucous membranes. Acute inhalation exposure symptoms include severe irritation and/or burns of the nose, throat, and respiratory tract, dyspnea (difficulty breathing), wheezing, chest pain, bronchospasm, foamy sputum, pulmonary edema, and even respiratory arrest. Contact with ammonia gas can cause skin irritation, corrosive burns, and blistering, while contact with liquid ammonia may result in frostbite and caustic burns. Eye exposure can lead to severe irritation, tearing, eye burns, and temporary or permanent damage. Ingestion may result in burns, corrosion, severe pain in the mouth, throat, esophagus, and stomach, and even death.

The following aid measures should be taken for individuals exposed to ammonia. In case of accidental inhalation, the person should move to a well-ventilated area and seek medical attention. In case of skin or eye contact, the affected area should be rinsed with plenty of water for at least 20 minutes.

Although ammonia is generally stable under normal conditions, it can react violently with certain substances. For example, ammonia reacts explosively with hydrocarbon gases, chlorine, and fluorine, as well as metals such as mercury and silver. Anhydrous ammonia should not come into contact with compounds such as acrolein, boron, chloric acid, chlorine monoxide, chlorites, perchlorates, sulfur, tin, and strong acids, as these reactions can be violent.

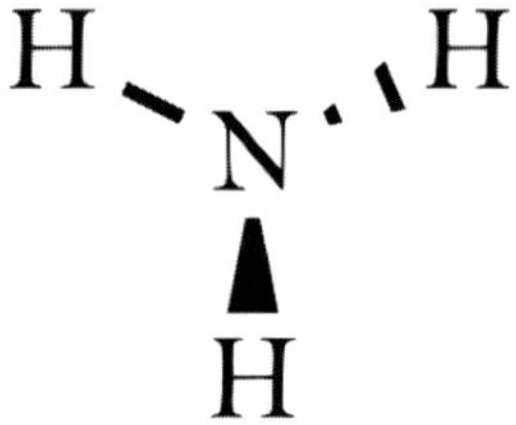 H H N H	**AMMONIA**	
	Other names: Anhydrous ammonia, nitrogen trihydride, azane, and spirit of hartshorn	
	Origin: Natural/synthetic	Formula: NH_3
Type: Inorganic	Discoverer: Joseph Priestley in 1773	MW: 17.03 g/mol
Appearance: colorless gas with a penetrating odor	Inventor: NA	mp: -77.7°C
Uses: Reagent, fertilizer, and energy reservoir	Toxicity: Yes	bp: -33.3°C

The Characterization of Ammonia

The molecular geometry of ammonia is a trigonal pyramid, and the electronic geometry is tetrahedral with the lone pair oriented at one vertex of the tetrahedron, and the N atom has sp^3 hybridization and oxidation number (3+). Various theoretical-experimental studies determine the exact structure of ammonia by providing reliable information on bond distances, ionization energy, electron density, and so on.

Spectroscopy techniques examine the elements of this small molecule, although because it is a gas, part of the characterization is indirect. The ^{1}H-NMR technique allows ammonia to be detected in a DMSO-d6, where the NH_4^+ ion (from NH_4Cl) appears as a triplet at approximately 6.9 ppm (Kolen et al., 2021). The concentration of NH_4^+ can be accurately determined by UV-vis spectrophotometry at 543 nm by indirectly measuring oxidation with potassium bromate ($KBrO_3$). This type of analysis is most suitable for determining very dilute concentrations, such as those required for trace analysis in water (Jiang & Tao, 2023).

Direct characterization of ammonia in the gas phase by mass spectrometry (NIST, 2025) allows the identification of the molecular ion and base peak at m/z = 17. While in the IR spectrum of ammonia gas, there are three intense and broad bands at 3337 cm^{-1}, 1628 cm^{-1}, and 968 cm^{-1} corresponding to the NH stretching and bending bonds, respectively (Preuss & Bechstedt, 2006).

Ammonia has also been characterized in the solid state using low-temperature X-ray diffraction crystallography at 178 K (Boese et al., 1997). Crystal refinement indicates that ammonia in the solid state belongs to space group 198, $P2_13$, with cell parameters a = 5.1305 Å, b = 5.1305 Å, and c = 5.1305 Å, and with angles $\alpha = 90°$, $\beta = 90°$, and $\gamma = 90°$. The existence of small bending vibrations within the NH_3 tetrahedron is determined in XRD and confirmed by the *ab initio* calculations. The crystal structure of ammonia is described as a cubic unit cell with three NH bonds of identical distance of 1.020 Å and with an HNH dihedral angle of 107.6° (Preuss & Bechstedt, 2006). Experimental measurements (μ exp = 1.471 D) and computational calculations (μ theo = 1.470 D) indicate that ammonia has a permanent dipole moment. Furthermore, the ammonia molecule is found as an L-type neutral ligand in many inorganic crystals, assessing its electron density donor capacity and its role in the geometry of the complex (Buchhorn & Krewald, 2023).

The Reactivity of Ammonia in Organic Synthesis

Ammonia is a Lewis base, nucleophile, and reducing agent involved in the synthesis of *N*-compounds such as amines, amides, and heterocycles. Reactions with ammonia can be carried out in the gas phase, in solution, and even in liquid ammonia at low temperatures. In addition, ammonia is a ligand used in inorganic chemistry.

Ammonia, as a Lewis base, reacts with Brønwsted-Lowry acids to form ammonium salts (Ouellette & Rawn, 2014). For example, with hydrochloric acid, ammonia forms ammonium chloride (Figure 8.1); with sulfuric acid, it forms ammonium sulfate (Figure 8.2); and with nitric acid, it forms ammonium nitrate (Figure 8.3). Some of these reactions are reviewed below.

As a reductant, ammonia reacts with oxidants such as oxygen. For example, the combustion of ammonia with pure oxygen is exothermic and forms nitrogen and water as products (Figure 8.4), but incomplete combustion also forms NOx species, mostly, nitrogen monoxide (NO) and nitrogen dioxide (NO_2) (F. Ma et al., 2023).

$$(1)\quad NH_3 + HCl(ac) \longrightarrow NH_4Cl$$

$$(2)\quad 2NH_3 + H_2SO_4(ac) \longrightarrow (NH_4)_2SO_4$$

$$(3)\quad NH_3 + HNO_3(ac) \longrightarrow NH_4NO_3$$

$$(4)\quad 4NH_3 + 3O_2 \xrightarrow{\Delta T} 2N_2 + 6H_2O$$

Figure 8. Reaction schemes to obtain ammonium salts (1-3) and combustion of ammonia (4).

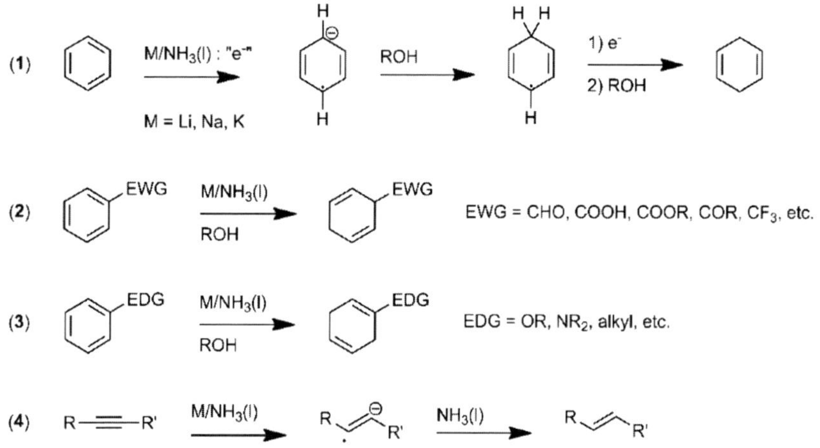

Figure 9. Ammonia-based reaction schemes for cyclohexadiene (1-3) and *trans* alkenes (4).

Liquid ammonia is used in reduction (hydrogenation) reactions of arenes (Zimmerman, 2012); for example, the Birch reduction of benzene produces 1,4-cyclohexadiene (Figure 9.1). In this reaction, liquid ammonia (-78°C) serves as the reaction medium and reagent. An alkali metal (Li, Na, or K) and ammonia react to produce an electride salt (e.g., $[Na(NH_3)_6]^+\ e^-$), which initiates the reduction via an ion-radical mechanism. The alcohol donates a proton, completing the ring's semihydrogenation and yielding the corresponding diene. The Birch reduction is regioselective, as the major product depends on the type of substituent, whether it is an electron-withdrawing group (EWG) (Figure 9.2) or an electron-donating group (EDG) (Figure 9.3). Phenols and halobenzenes do not undergo the Birch reaction.

Similarly, liquid ammonia reduces alkynes to trans-alkenes (Benkeser & Belmonte, 1984). In these reactions, the vinyl anion, an intermediate species, is sufficiently basic to extract a proton from ammonia, yielding the corresponding semihydrogenation product (Figure 9.4).

Ammonia, as a nucleophile, reacts with primary alkyl halides (RX) to form primary, secondary, and tertiary amines (Afanasyev et al., 2021). Among these, alkyl iodides (RI) and bromides (RBr) have better leaving groups, enabling milder reaction conditions, whereas alkyl chlorides (RCl) usually require longer reaction times and higher temperatures.

Ammonia substitutes the X group via an S_N2 mechanism to form the primary amine RNH_2 (Figure 10.1). However, despite a stoichiometric reaction, a mixture of primary (monosubstituted), secondary (disubstituted), and tertiary (trisubstituted) amine products is still formed (Figure 10.2). Primary amines are more nucleophilic than ammonia, and secondary amines are, in turn, more nucleophilic than primary amines (Ju & Varma, 2004). Therefore, one way to increase the yield of the primary amine is to use an excess of ammonia. Nevertheless, a mixture of amines is obtained because tertiary amines are favored due to their increased nucleophilicity and reactivity.

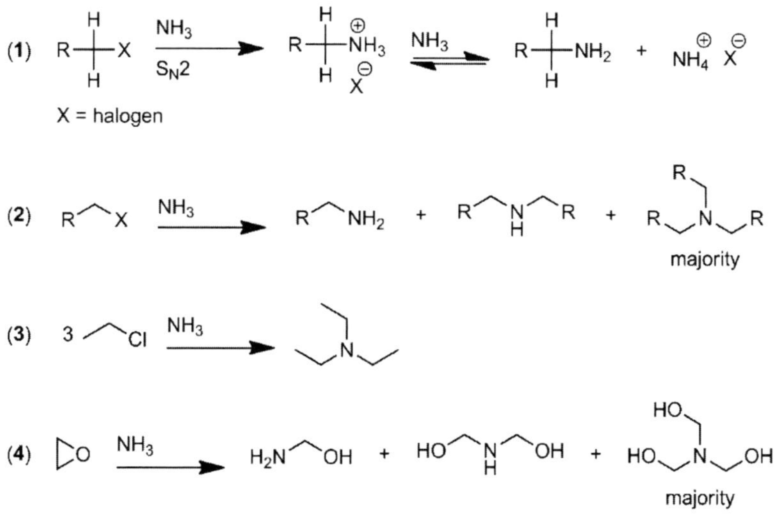

Figure 10. S_N2 reaction mechanism of a primary halide with ammonia (1) and reaction examples (2–3), as well as epoxy ring opening (4).

The reaction of ammonia with three equivalents of a halide or with an excess of halide favors the formation of tertiary amines as the major product. For example, the reaction of three equivalents of chloroethane with one equivalent of ammonia produces triethylamine (Figure 10.3). Alternative methods for synthesizing triethylamine include the use of ethanol with an osmium catalyst or ethylene with an iron catalyst (Afanasyev et al., 2021); however, diethylamine is also produced during the process.

Ammonia reacts with ethylene oxide via an S_N2 mechanism, producing a mixture of ethanolamine, diethanolamine, and triethanolamine (Figure 10.4) (Devaraja & Kiss, 2022). Among these, triethanolamine is the favored product (Andreev et al., 2015).

Ammonia is mainly involved in condensation reactions with carbonyls; for example, ammonia in the gas phase is condensed with carbon dioxide (CO_2) under high pressure and temperature conditions to form the carbamate intermediate and urea as the final product (Figure 11.1), as described in the original Bosch-Meiser process (Zhao et al., 2023). Ammonia reacts with other carbonyl compounds such as acyl halides (RCOX), anhydrides (RCOOCOR), esters (RCOOR), and carboxylic acids (RCOOH) to produce primary amides (Figure 11.2) in the following order of reactivity: halogen > OCOR > OR > OH (Norman & Coxon, 2017). Ammonia also reacts with aldehydes and certain ketones containing an alpha hydrogen to produce Schiff bases as intermediates. Reductive amination (Figure 11.3) occurs in the presence of a reducing agent, such as sodium borohydride ($NaBH_4$), lithium aluminum hydride ($LiAlH_4$), or the catalyst Raney nickel with H_2, producing the corresponding primary amine (Ainsworth, 1956; Bäumler et al., 2020).

The reactivity of ammonia extends to the synthesis of other condensation products (Norman & Coxon, 2017), for example, ammonia and nitriles produce amidines (Figure 12.1), with cyanate or isocyanate, urea is also synthesized (Figure 12.2), and with thiocyanate or thioisocyanate, thiourea is formed (Figure 12.3).

Ammonia participates in a variety of reactions with arenes, most notably in nucleophilic aromatic substitution (S_NAr) reactions of aryl halides catalyzed by Cu and Pd (Figure 13.1) (Aubin et al., 2010). However, the reaction conditions influence the number of products formed.

For example, when chlorobenzene reacts with excess ammonia (gas phase), and using a Cu catalyst at 450°C, three products are formed via a benzyne intermediate mechanism (Figure 13.2). The reduction product is benzene, the S_NAr product is aniline, and the second substitution product (double substitution) is diphenylamine (Burgers & Vanbekkum, 1994).

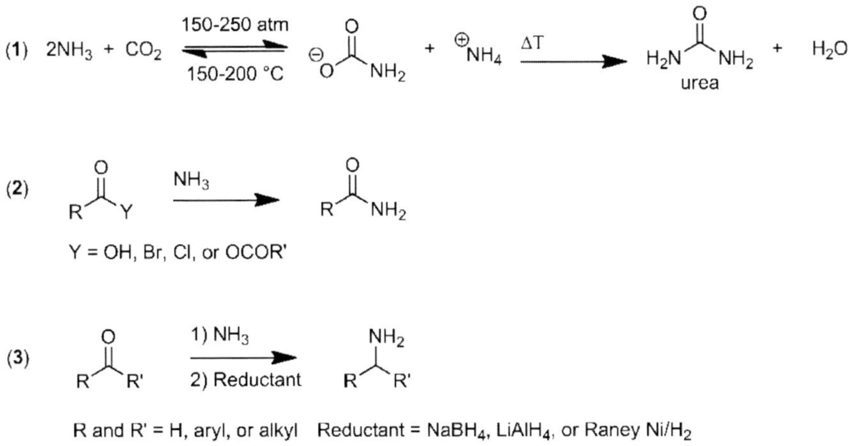

(1) $2NH_3 + CO_2 \underset{150\text{-}200\ °C}{\overset{150\text{-}250\ atm}{\rightleftharpoons}} {}^{\ominus}O\text{-}C(=O)\text{-}NH_2 + {}^{\oplus}NH_4 \xrightarrow{\Delta T} H_2N\text{-}C(=O)\text{-}NH_2\ (\text{urea}) + H_2O$

(2) $R\text{-}C(=O)\text{-}Y \xrightarrow{NH_3} R\text{-}C(=O)\text{-}NH_2$

Y = OH, Br, Cl, or OCOR'

(3) $R\text{-}C(=O)\text{-}R' \xrightarrow[\text{2) Reductant}]{\text{1) } NH_3} R\text{-}CH(NH_2)\text{-}R'$

R and R' = H, aryl, or alkyl Reductant = $NaBH_4$, $LiAlH_4$, or Raney Ni/H_2

Figure 11. Reaction schemes of carbonyl condensation with ammonia (1, 2) and reductive amination (3).

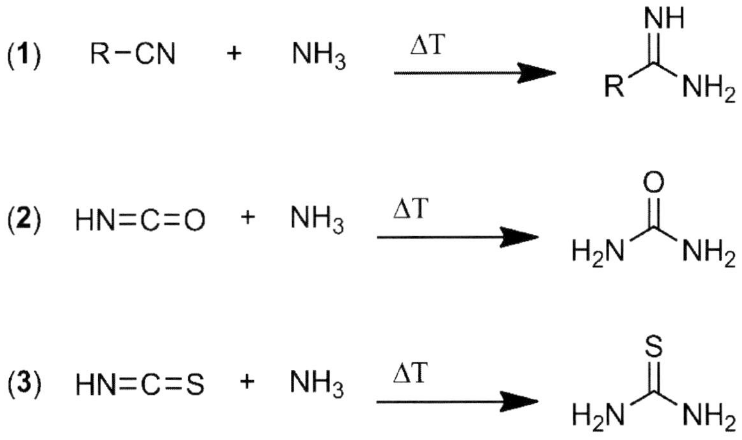

(1) $R\text{-}CN + NH_3 \xrightarrow{\Delta T} R\text{-}C(=NH)\text{-}NH_2$

(2) $HN{=}C{=}O + NH_3 \xrightarrow{\Delta T} H_2N\text{-}C(=O)\text{-}NH_2$

(3) $HN{=}C{=}S + NH_3 \xrightarrow{\Delta T} H_2N\text{-}C(=S)\text{-}NH_2$

Figure 12. Reaction schemes for synthesizing urea (1, 2) and thiourea (3).

In aromatic rings with electron-withdrawing groups (EWGs), substitution does not always require high temperatures. For example, 2,4-difluoronitrobenzene (Figure 13.3) reacts with ammonia at 25°C, primarily replacing fluorine in the ortho position (Ji et al., 2011).

S_NAr aminations with ammonia in an aqueous medium can also occur in some heterocyclic rings under mild conditions with Cu(I) catalysts (Figure 13.4). For example, the reaction of ammonia with halopyridines yields aminopyridines at lower temperatures than those required for phenyl rings (Elmkaddem et al., 2010; Schranck & Tlili, 2018).

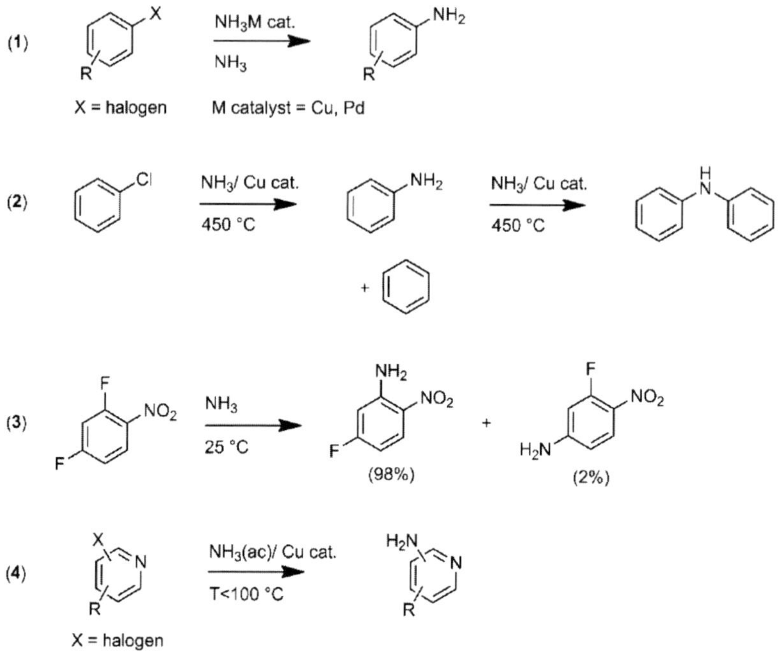

Figure 13. S_NAr reaction schemes with ammonia.

Ammonia is involved not only in S_NAr with heterocycles, but also in the synthesis of n-membered rings such as pyrrolidines, piperidines, pyrroles, and pyridines, among others (Wiley et al., 1962). The following are examples of ammonia used as a reagent in pyridine assembly reactions.

Among the most popular methods for the synthesis of pyridines is the Chichibabin reaction, where three equivalents of an aldehyde, ketone, α-β-unsaturated carbonyl, or combinations of them react with ammonia to form alkyl-substituted pyridines, but yields are low often (Frank & Seven, 1949).

The use of zeolite catalysts increases yields; for example, the reaction of formaldehyde, acetaldehyde, and ammonia produces pyridine with up to 60% yield (Figure 14.1) (Shimizu et al., 1998). Methyl pyridines can also be synthesized from acetylene and ammonia (Akhmerov et al., 1975), yielding two structural isomers with the methyl group in the ortho and para positions (Figure 14.2).

Finally, the Hantzsch pyridine synthesis (Katritzky et al., 2010) involves the condensation of an aldehyde, two equivalents of a β-ketoester, and one equivalent of ammonia, forming a dihydropyridine intermediate that is subsequently treated to yield the substituted pyridine (Figure 14.3).

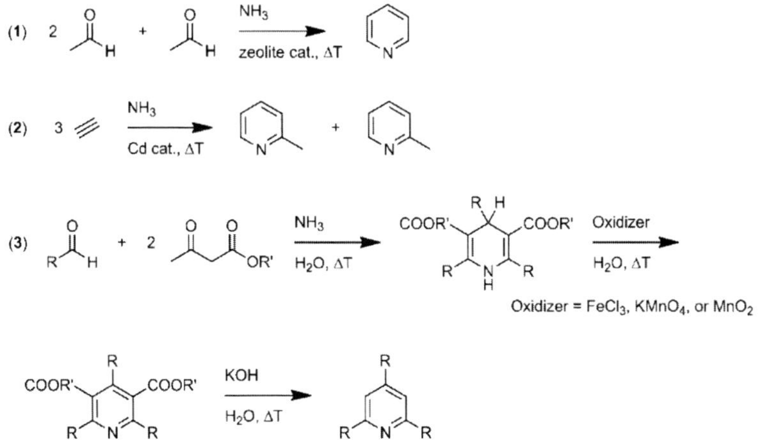

Figure 14. Reaction schemes for the synthesis of pyridines from ammonia and carbonyls.

Ammonia is crucial for the synthesis of α-amino acids, which are essential for biochemistry, biology, and all living organisms (Norman & Coxon, 2017). For example, ammonia reacts with α-bromocarboxylic acids in a substitution reaction to produce the corresponding α-amino acid (Figure 15.1).

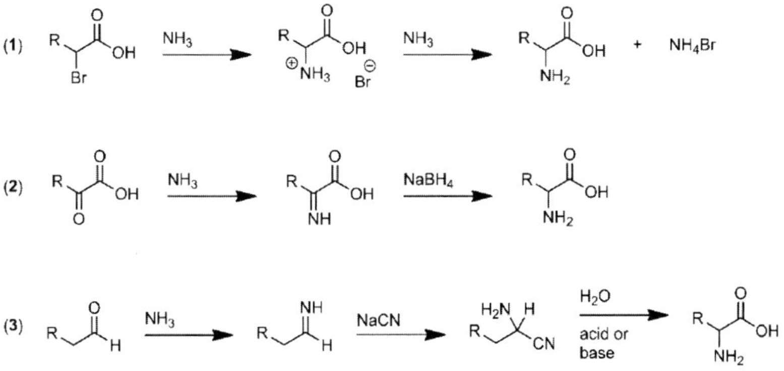

Figure 15. Reaction schemes for the synthesis of α-amino acids from ammonia and carboxylic acids.

Another method for obtaining α-amino acids is the Gabriel procedure, which involves the reductive amination of α-keto acids with ammonia, followed by reduction with $NaBH_4$ (Figure 15.2).

Finally, the Strecker synthesis is a three-component reaction of ammonia, cyanide, and an aldehyde. In the first step, ammonia and cyanide attack the electrophilic carbonyl, forming the intermediate α-aminonitrile, which is then hydrolyzed with acid or base in an aqueous medium to yield the corresponding α-amino acid (Figure 15.3).

In the above-mentioned syntheses, the α-amino acids produce a racemic mixture of L- and R-isomers that must be separated in subsequent steps—a process known as amino acid resolution. This process employs various techniques, such as forming salts or derivatives that are separated by differences in solubility or using chromatography to separate isomers based on their retention times in the stationary phase.

Another strategy is the asymmetric synthesis of amino acids, which involves identifying reaction conditions, catalysts, or reagents (e.g., enzymes) to produce an enantiomeric excess and eliminate the need for racemic resolution (Barrett, 1985).

The Reactivity of Ammonia as a Ligand

In inorganic chemistry, ammonia is a neutral (L-type) monodentate σ-donor that participates in reactions with metal centers to form stable complexes. Additionally, ammonia undergoes redox reactions with alkali metals such as Li, Na, or K (Jain et al., 2017), forming the corresponding amide or azanide (MNH_2) (Figure 16.1), in which the metal center increases its oxidation state.

Ammonia reacts with some metal oxides. For example, the synthesis of the Tollens reagent (used for the identification of aldehydes) involves the reaction of silver oxide (Ag_2O) with ammonia in an aqueous medium to form diamine silver(I) hydroxide, $[Ag(NH_3)_2]OH$, which is the active complex in the silver mirror test (Figure 16.2) and oxidizes aldehydes to carboxylic acids (Fappiano et al., 2022).

Ammonia participates in addition and substitution reactions with transition metal salts MX_m. The addition products have the structure $M(NH_3)_nX_m$, while the substitution compounds have the structure $M(NH_3)_nX_{m-n}$. Ammonia forms complexes with practically all transition elements. The most common reports are with metal centers Cr, Mn, Fe, Co,

Ni, Zn, and Cu. A classic example is the amino copper complexes $[Cu(NH_3)_n(H_2O)_{m-n}]^{2+}$, where ammonia (ligand ammine) successively displaces water (ligand aqua) (Anufrienko et al., 2012). In aqueous Cu(II) salts (e.g., CuX_2), water ligands in $[Cu(H_2O)_6]^{2+}$ are replaced by ammonia, forming the dark blue complex $[Cu(NH_3)_4(H_2O)_2]^{2+}$. Adding six equivalents of ammonia replaces all water ligands, forming $[Cu(NH_3)_6]^{2+}$, which exhibits a violet-indigo color (Figure 16.3).

Ammonia chemistry with platinum group metals has applications in reductive catalytic reactions (Möller & Meyer, 1992) and in amination reactions of aryl halides with Pd(II) and Cu(I) catalysts (Aubin et al., 2010). Moreover, the role of ammonia is even more relevant since it is involved in the synthesis of chemotherapy drugs (Dasari & Bernard Tchounwou, 2014). For example, the synthesis of cisplatin (Figure 16.4) begins with a Pt(II) salt, usually potassium tetrachloroplatinate (K_2PtCl_4), which reacts with potassium iodide (KI) to form the tetraiodoplatinate anion $[PtI_4]^{2-}$. This reaction is followed by the addition of two equivalents of ammonia, replacing the two iodide leaving groups to produce cis-$[Pt(NH_3)_2I_2]$. In the next two steps, the iodine ligands are replaced by chlorine ligands using $AgNO_3$ and potassium chloride (KCl) to form the cisplatin complex cis-$[Pt(NH_3)_2Cl_2]$, which is the biologically active compound (Wilson & Lippard, 2014).

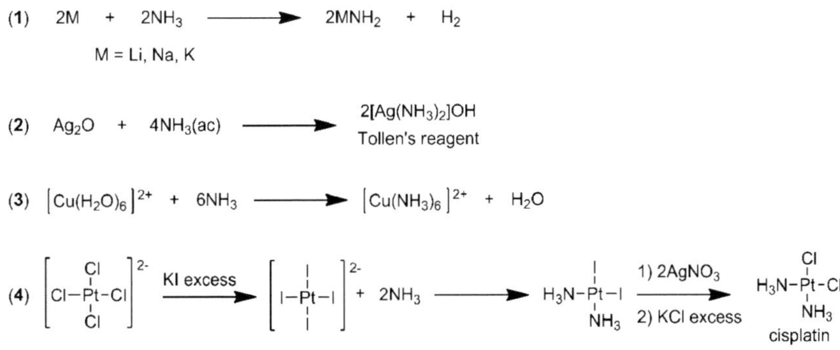

(1) $2M + 2NH_3 \longrightarrow 2MNH_2 + H_2$

M = Li, Na, K

(2) $Ag_2O + 4NH_3(ac) \longrightarrow 2[Ag(NH_3)_2]OH$

Tollen's reagent

(3) $[Cu(H_2O)_6]^{2+} + 6NH_3 \longrightarrow [Cu(NH_3)_6]^{2+} + H_2O$

(4) $[PtCl_4]^{2-} \xrightarrow{\text{KI excess}} [PtI_4]^{2-} + 2NH_3 \longrightarrow H_3N\text{-}Pt(I)(NH_3)\text{-}I \xrightarrow[\text{2) KCl excess}]{\text{1) 2AgNO}_3} H_3N\text{-}Pt(Cl)(NH_3)\text{-}Cl$

cisplatin

Figure 16. Examples of ammonia's reactivity with alkali metals (1), Ag_2O (2), Cu(II) (3), and the synthesis of cisplatin (4).

The Ammonia Industry

China remained the world leader in anhydrous ammonia production, generating approximately 30% of the global supply in 2024 (IEA, 2025). Global ammonia production increased from 131 million metric tons to 150 million metric tons over the last decade, projecting an upward trend (Statista, 2025b). In 2023, China was the leading producer, with an output of 43 million metric tons, followed by Russia, India, and the USA, each producing 14 million metric tons (Erdemir & Dincer, 2024). Global ammonia production capacity is projected to reach 289.8 million metric tons by 2030 (Statista, 2025a). According to 2023 reports (Fortune Business Insights, 2025), the leading ammonia producers, in order, were Yara International, BASF, CF Industries Holdings, Nutrien, SABIC, Koch Fertilizer, Praxair, Rashtriya Chemicals and Fertilizers Limited, Qatar Fertilizer Company, and OCI Nitrogen.

Total ammonia emissions in 2022 were estimated to have exceeded 64 million metric tons worldwide. By sector, agriculture was the largest contributor to ammonia emissions in 2022, generating 49.43 million metric tons, followed by the waste sector (9.03 million metric tons) and the energy sector (3.04 million metric tons) (Our World in Data, 2025).

What Would Happen If Synthetic Ammonia Production Were Stopped?

No substitute for ammonia can efficiently replace the wide range of products derived from it, particularly at the scale required to meet the demands of the current global population. Although certain ammonia derivatives, such as ammonium chloride or urea, can be used as alternatives in organic synthesis routes, these nitrogen precursors are themselves derived from synthetic ammonia, rendering them unviable as true substitutes. A possible alternative source would be the extraction of ammonia-rich mineral deposits from ammonium salts (Ridgway et al., 1990). However, this option has significant drawbacks, such as the scarcity of ammonium deposits. Additionally, ammonia minerals are non-renewable resources, and their extraction could harm the environment. Furthermore, producing anhydrous ammonia through alternative methods would be more expensive.

If ammonia production were to decline significantly or stop entirely, both the global economy and the food production chain would face severe consequences. These effects would be particularly immediate in urban areas. First, if ammonia production were to fall significantly, shortages would raise the cost of imported raw materials. In the long term, minimum food requirements would go unmet. Moreover, if global ammonia production were to cease entirely, the absence of ammonia-derived fertilizers would significantly raise the cost of producing and shipping food, resulting in a permanent increase in food prices. This scarcity of fertilizers could lead to uncontrolled inflation and disproportionately impact poorer countries with limited natural resources (Tonelli et al., 2024).

There is substantial evidence supporting the success of the Bosch-Haber process. After the initiation of synthetic ammonia production in 1913, the global population increased from approximately 1.65 billion in 1900 to 6.8 billion in 2008—an unprecedented growth in human history. This was not a coincidence; in fact, it is estimated that the percentage of the population dependent on food produced with synthetic ammonia fertilizers increased from 42% in 2000 to 48% in 2008 (Ritchie, 2025). Without this technological innovation, the global population in 2000 would have been approximately 3.5 to 4 billion, rather than 6.1 billion in 2000 and 8.2 billion in 2024.

An Opinion on Ammonia

Ammonia was among the first synthetic small molecules produced in the last century for practical applications. Molecules like ammonia largely drove the rapid progress of organic chemistry in the 20th century. These advancements have enabled the development of next-generation medicines, fertilizers, and fuel energy reservoirs (Q. Ma et al., 2025).

Ammonia is a chemical precursor of important compounds such as urea, ammonium nitrate, and ammonium hydroxide. These compounds are essential to produce other materials and compounds used in the agricultural, processed food, pharmaceutical, and chemical industries in general. Studies estimate that the agricultural sector consumes between 70% and 90% of global ammonia production for use as synthetic fertilizers (Wang et al., 2024).

One major drawback of ammonia production is that obtaining hydrogen from natural gas generates significant volumes of carbon dioxide. For every two molecules of ammonia produced, at least three molecules of carbon dioxide are released into the atmosphere (Figure 7.1 and Figure 7.2). An alternative method, using coal to produce hydrogen gas, is also environmentally unsustainable. The exploitation of coal deposits often causes ecological damage and widespread environmental pollution, as is frequently observed in the mining industry (Holley & Shearing, 2017; Short & Crook, 2022). This represents a significant global challenge, as current ammonia production processes contribute to the accumulation of greenhouse gases, which drive annual increases in Earth's average temperature (Filonchyk et al., 2024). Furthermore, elevated concentrations of ammonia gas affect air quality, posing health risks to humans (R. Ma et al., 2021) and other animals (Bleizgys & Naujokienė, 2023).

Another significant concern regarding ammonia is its use in military applications. Some ammonia derivatives, such as ammonium nitrate, have long been used in explosive formulations (Akhavan, 2022) and propellants (Oommen & Jain, 1999). Other derivatives, such as nitrogen mustards, have been employed as chemical weapons in warfare (Lombardo, 2017). The use of technology to develop weapons for hostile purposes is an ethically concerning issue. Given the significant need to improve the quality of life for living beings, resources should be directed toward constructive, rather than destructive, objectives.

The Importance of Ammonia

Ammonia is an inorganic compound that plays a critical role in organic synthesis and is essential to produce several raw materials. The chemistry of ammonia-derived compounds is complex, but it can be summarized in fertilizers, medicines, and cleaning products (Erisman et al., 2008).

Ammonia is the most cost-effective source of chemically active nitrogen. As a result, fertilizers such as urea and ammonium nitrate can be synthesized at relatively low costs. Fertilizers transform unproductive land into arable and profitable farmland, thereby increasing food availability for large populations. Abundant food, in turn, helps alleviate the persistent famine experienced by a significant portion of the global population (Sheeran, 2011). Thus, the current state of the global food sector is closely tied to ammonia production as a raw material for synthetic fertilizers and its associated applications (Zarebska et al., 2015).

Moreover, the value of ammonia may rise in the coming decades due to its potential as a fuel. When ammonia reacts with oxygen, it releases significant heat energy ($\Delta H = -1267.2$ kJ/mol), making the combustion reaction highly exothermic. Pure ammonia can be used in propulsion systems or as a cleaner alternative to fossil fuels, as its primary combustion products—nitrogen and water—are environmentally harmless. However, a major challenge lies in achieving efficient and safe ignition, necessitating the development of specialized ammonia combustion systems (Valera-Medina et al., 2021). Recent research has explored mixtures of diesel, natural gas, and ammonia to get green fuels with high energy output and scalability (Jayabal, 2024).

If science can overcome the challenges related to ammonia storage, transportation, and combustion efficiency, green ammonia—particularly—will become one of the cleaner fuels for future generations (Li et al., 2021).

Some Curiosities About Ammonia

- The pseudo-Geber treatise *De Inventione Veritatis,* written between the 13th and 14th centuries (Reti, 1965), describes a method for producing aqua regia using ammonium chloride and nitric acid. Aqua regia, an acidic mixture capable of dissolving gold, is still used today in gold purification methods and in research involving chemistry and materials science.
- In the 17th century, ammonium carbonate was produced by heating and distilling the horns and hooves of deer (commonly referred to as hart) and other animals. The resulting liquid, known as the spirit of hartshorn, was used as a stimulant to revive individuals from fainting and as a detergent, among other applications (Freemantle, 2016).
- The chemist Michele Peyrone (1813–1883) first synthesized cisplatin in 1844 (Kauffman et al., 2010). More than a century later, its cytotoxic activity, which selectively targets cancer cells, was discovered. Since 1978, cisplatin has been included in the World Health Organization's list of essential medicines (World Health Organization, 2025). It is currently used to treat a variety of cancers, including cervical cancer, head and neck cancer (as a radiosensitizer), low-grade glioma, nasopharyngeal cancer (as a radiosensitizer), non-small cell lung cancer, osteosarcoma, ovarian germ cell tumors, and testicular germ cell tumors.
- The Oppau plant (now part of Ludwigshafen) in Germany exploded on September 21, 1921, when a silo containing approximately 4,500

tons of ammonium nitrate and sulfate was detonated with dynamite—a routine practice at the time. The explosion resulted in the deaths of over 500 people and injuries to more than 2,000 others ("The Oppau Explosion," 1921).

- On April 16, 1947, the French ship *SS Grandcamp,* carrying chemicals including ammonium nitrate, exploded in Galveston Bay, Texas. The detonation triggered a series of fires and explosions throughout the city, killing an estimated 581 people (Stephens, 1997). Decades later, on April 17, 2013, an explosion at the West Fertilizer Company in Texas, caused by stored ammonium nitrate, resulted in 15 deaths and 260 injuries (Babrauskas, 2018).
- In a 2016 study, ammonia was detected for the first time in the troposphere (12–15 km above the Earth's surface). This phenomenon was attributed to global warming, intensified agriculture, and increased fertilizer use (Höpfner et al., 2016).
- On August 4, 2020, a chemical detonation occurred in Beirut, Lebanon, when an uncontrolled fire reached a warehouse storing approximately 2,750 tons of ammonium nitrate. The resulting explosion killed 220 people and injured over 6,500 others (Al-Hajj et al., 2021).
- Around 180 million tons of urea were produced in 2024 (Devkota et al., 2024). Nearly all of this production relies on synthetic ammonia as a feedstock, with most urea used as fertilizer.
- Ammonia is a carbon-free energy reservoir because it emits only nitrogen and water, making it a strong candidate for green fuel (Jafar et al., 2024).

References

Afanasyev, O. I., Kuchuk, E. A., Muratov, K. M., Denisov, G. L., & Chusov, D. (2021). Symmetrical tertiary amines: Applications and synthetic approaches. *European Journal of Organic Chemistry*, *2021*(4), 543–586. https://doi.org/10.1002/ejoc.202001171.

Ainsworth, C. (1956). The reductive alkylation of primary aromatic amines with Raney nickel and alcohols. *Journal of the American Chemical Society*, *78*(8), 1635–1636. https://doi.org/10.1021/ja01589a036.

Akhavan, J. (2022). Introduction to explosives. In J. Akhavan (Ed.), *The Chemistry of Explosives* (pp. 1–27). The Royal Society of Chemistry. https://doi.org/10.1039/BK9781839164460-00001.

Akhmerov, K. M., Yusupov, D., Abdurakhmanov, A., & Kuchkarov, A. B. (1975). Catalytic synthesis of pyridine and methylpyridine from acetylene and ammonia. *Chemistry of Heterocyclic Compounds*, *11*(2), 190–192. https://doi.org/10.1007/BF00471395.

Al-Hajj, S., Dhaini, H. R., Mondello, S., Kaafarani, H., Kobeissy, F., & DePalma, R. G. (2021). Beirut ammonium nitrate blast: Analysis, review, and recommendations. *Frontiers in Public Health*, *9*, 1–11. https://doi.org/10.3389/fpubh.2021.657996.

Andreev, D. V., Makarshin, L. L., Gribovskii, A. G., Yushchenko, D. Yu., Sergeev, E. E., Zhizhina, E. G., Pai, Z. P., & Parmon, V. N. (2015). Triethanolamine synthesis in a continuous flow microchannel reactor. *Chemical Engineering Journal*, *259*, 252–256. https://doi.org/10.1016/j.cej.2014.07.118.

Anufrienko, V. F., Shutilov, R. A., Zenkovets, G. A., Gavrilov, V. Yu., Vasenin, N. T., Shubin, A. A., Larina, T. V., Zhuzhgov, A. V., Ismagilov, Z. R., & Parmon, V. N. (2012). The state of Cu2+ ions in concentrated aqueous ammonia solutions of copper nitrate. *Russian Journal of Inorganic Chemistry*, *57*(9), 1285–1290. https://doi.org/10.1134/S0036023612090033.

Aubin, Y., Fischmeister, C., Thomas, C. M., & Renaud, J.-L. (2010). Direct amination of aryl halides with ammonia. *Chemical Society Reviews*, *39*(11), 4130–4145. https://doi.org/10.1039/c003692g.

Babrauskas, V. (2018). The ammonium nitrate explosion at West, Texas: A disaster that could have been avoided. *Fire and Materials*, *42*(2), 164–172. https://doi.org/10.1002/fam.2468.

Barrett, G. C. (1985). Resolution of amino acids. In G. C. Barrett (Ed.), *Chemistry and Biochemistry of the Amino Acids* (1st ed., pp. 338–353). Springer Netherlands. https://doi.org/10.1007/978-94-009-4832-7_10.

Bäumler, C., Bauer, C., & Kempe, R. (2020). The synthesis of primary amines through reductive amination employing an iron catalyst. *ChemSusChem*, *13*(12), 3110–3114. https://doi.org/10.1002/cssc.202000856.

Benkeser, R. A., & Belmonte, F. G. (1984). Reduction of alkynes by a new reducing system. *The Journal of Organic Chemistry*, *49*(9), 1662–1664. https://doi.org/10.1021/jo00183a038.

Berl, E. (1937). Fritz Haber. *Journal of Chemical Education*, *14*(5), 203–207. https://doi.org/10.1021/ed014p203.

Berwal, P., Kumar, S., & Khandelwal, B. (2021). A comprehensive review on synthesis, chemical kinetics, and practical application of ammonia as future fuel for combustion. *Journal of the Energy Institute*, *99*, 273–298. https://doi.org/10.1016/j.joei.2021.10.001.

Bleizgys, R., & Naujokienė, V. (2023). Ammonia emissions from cattle manure under variable moisture exchange between the manure and the environment. *Agronomy*, *13*(6), 1555. https://doi.org/10.3390/agronomy13061555.

Boese, R., Niederprüm, N., Bläser, D., Maulitz, A., Antipin, M. Yu., & Mallinson, P. R. (1997). Single-crystal structure and electron density distribution of ammonia at 160 K on the basis of X-ray diffraction data. *The Journal of Physical Chemistry B*, *101*(30), 5794–5799. https://doi.org/10.1021/jp970580v.

Bosch, C. (1908). *Process of producing ammonia* (Patent US990191A).

Bosch, C., & Mittasch, A. (1911). *Production of ammonia* (Patent US1158167A).

Bosch, C., Mittasch, A., Wolf, H., & Stern, G. (1910). *Catalytic agent for use in producing ammonia* (Patent US1148570A).

Buchhorn, M., & Krewald, V. (2023). The π-interactions of ammonia ligands evaluated by *ab initio* ligand field theory. *Dalton Transactions*, *52*(20), 6685–6692. https://doi.org/10.1039/D3DT00511A.

Burgers, M. H. W., & Vanbekkum, H. (1994). Vapor-phase substitution of chlorobenzene with ammonia, catalyzed by copper-exchanged zeolites. *Journal of Catalysis*, *148*(1), 68–75. https://doi.org/10.1006/jcat.1994.1186.

Cameo Chemicals. (2025, January 8). *Ammonia, Anhydrous*. Https://Cameochemicals.Noaa.Gov/Chemical/4860.

Dasari, S., & Bernard Tchounwou, P. (2014). Cisplatin in cancer therapy: Molecular mechanisms of action. *European Journal of Pharmacology*, *740*, 364–378. https://doi.org/10.1016/j.ejphar.2014.07.025.

Dellagi, A., Quillere, I., & Hirel, B. (2020). Beneficial soil-borne bacteria and fungi: A promising way to improve plant nitrogen acquisition. *Journal of Experimental Botany*, *71*(15), 4469–4479. https://doi.org/10.1093/jxb/eraa112.

Devaraja, D., & Kiss, A. A. (2022). Novel intensified process for ethanolamines production using reactive distillation and dividing-wall column technologies. *Chemical Engineering and Processing - Process Intensification*, *179*, 109073. https://doi.org/10.1016/j.cep.2022.109073.

Devkota, S., Karmacharya, P., Maharjan, S., Khatiwada, D., & Uprety, B. (2024). Decarbonizing urea: techno-economic and environmental analysis of a model hydroelectricity and carbon capture based green urea production. *Applied Energy*, *372*, 123789. https://doi.org/10.1016/j.apenergy.2024.123789.

Elmkaddem, M. K., Fischmeister, C., Thomas, C. M., & Renaud, J.-L. (2010). Efficient synthesis of aminopyridine derivatives by copper catalyzed amination reactions. *Chem. Commun.*, *46*(6), 925–927. https://doi.org/10.1039/B916569J.

Encyclopedia Mythica. (2025, January 8). *Ammon*. Pantheon.Org/Articles/a/Ammon.Html.

Erdemir, D., & Dincer, I. (2024). A quicker route to hydrogen economy with ammonia. *International Journal of Hydrogen Energy*, *82*, 1230–1237. https://doi.org/10.1016/j.ijhydene.2024.08.067.

Erisman, J. W., Sutton, M. A., Galloway, J., Klimont, Z., & Winiwarter, W. (2008). How a century of ammonia synthesis changed the world. *Nature Geoscience*, *1*(10), 636–639. https://doi.org/10.1038/ngeo325.

Fappiano, L., Carriera, F., Iannone, A., Notardonato, I., & Avino, P. (2022). A review on recent sensing methods for determining formaldehyde in agri-food chain: A comparison with the conventional analytical approaches. *Foods*, *11*(9), 1351. https://doi.org/10.3390/foods11091351.

Filonchyk, M., Peterson, M. P., Zhang, L., Hurynovich, V., & He, Y. (2024). Greenhouse gases emissions and global climate change: Examining the influence of CO2, CH4, and N2O. *Science of The Total Environment*, *935*, 173359. https://doi.org/10.1016/j.scitotenv.2024.173359.

Fortune Business Insights. (2025, January 8). *Top 10 Leading Ammonia Manufacturers in the Market, 2023.* Https://Www.Fortunebusinessinsights.Com/Blog/Top-Ammonia-Manufacturers-Worldwide-10922.

Francis, C. A., Beman, J. M., & Kuypers, M. M. M. (2007). New processes and players in the nitrogen cycle: The microbial ecology of anaerobic and archaeal ammonia oxidation. *The ISME Journal*, *1*(1), 19–27. https://doi.org/10.1038/ismej.2007.8.

Frank, R. L., & Seven, R. P. (1949). Pyridines. IV. A study of the Chichibabin synthesis. *Journal of the American Chemical Society*, *71*(8), 2629–2635. https://doi.org/10.1021/ja01176a008.

Freemantle, M. (2016, June 28). *Ammonium Carbonate*. Https://Www.Chemistryworld.Com/Podcasts/Ammonium-Carbonate/1010129.Article.

Haber, F., & Le Rossignol, R. (1909). *Production of ammonia* (Patent US1202995A).

Haber, F., & van Oordt, G. (1905). Über die bildung von ammoniak den elementen. *Zeitschrift Für Anorganische Chemie*, *44*(1), 341–378. https://doi.org/10.1002/zaac.19050440122.

Harper, D. (2025, January 8). *Etymology of ammonia*. Www.Etymonline.Com/Word/Ammonia.

Holley, C., & Shearing, C. (2017). *Criminology and the Anthropocene* (C. Holley & C. Shearing, Eds.; 1st ed.). Routledge. https://doi.org/10.4324/9781315541938.

Höpfner, M., Volkamer, R., Grabowski, U., Grutter, M., Orphal, J., Stiller, G., von Clarmann, T., & Wetzel, G. (2016). First detection of ammonia (NH3) in the Asian summer monsoon upper troposphere. *Atmospheric Chemistry and Physics*, *16*(22), 14357–14369. https://doi.org/10.5194/acp-16-14357-2016.

Humphreys, J., Lan, R., & Tao, S. (2021). Development and recent progress on ammonia synthesis catalysts for Haber–Bosch process. *Advanced Energy and Sustainability Research*, *2*(1), 1–23. https://doi.org/10.1002/aesr.202000043.

IEA. (2025, January 8). *Ammonia Technology Roadmap*. Https://Www.Iea.Org/Reports/Ammonia-Technology-Roadmap/Executive-Summary.

Jafar, U., Nuhu, U., Khan, W. U., & Hossain, M. M. (2024). A review on green ammonia as a potential CO2 free fuel. *International Journal of Hydrogen Energy*, *71*, 857–876. https://doi.org/10.1016/j.ijhydene.2024.05.128.

Jain, A., Kumar, S., Miyaoka, H., Zhang, T., Isobe, S., Ichikawa, T., & Kojima, Y. (2017). Ammonia suppression during decomposition of sodium amide by the addition of metal hydride. *International Journal of Hydrogen Energy*, *42*(35), 22388–22394. https://doi.org/10.1016/j.ijhydene.2017.02.049.

Jayabal, R. (2024). Ammonia as a potential green dual fuel in diesel engines: A review. *Process Safety and Environmental Protection*, *188*, 1346–1354. https://doi.org/10.1016/j.psep.2024.06.012.

Ji, P., Atherton, J. H., & Page, M. I. (2011). The kinetics and mechanisms of aromatic nucleophilic substitution reactions in liquid ammonia. *The Journal of Organic Chemistry*, *76*(9), 3286–3295. https://doi.org/10.1021/jo200170z.

Jiang, F., & Tao, J. (2023). UV-Visible spectrophotometric method for the determination of ammonia nitrogen using potassium bromate as oxidant. *Desalination and Water Treatment*, *306*, 159–164. https://doi.org/10.5004/dwt.2023.29778.

Johnson, B. (2022a). *Making Ammonia* (B. Johnson, Ed.; 1st ed.). Springer International Publishing. https://doi.org/10.1007/978-3-030-85532-1.

Johnson, B. (2022b). The state of ammonia synthesis at the turn of the twentieth century: The arena for discovery. In B. Johnson (Ed.), *Making Ammonia* (1st ed., pp. 77–89). Springer International Publishing. https://doi.org/10.1007/978-3-030-85532-1_9.

Ju, Y., & Varma, R. S. (2004). Aqueous N-alkylation of amines using alkyl halides: direct generation of tertiary amines under microwave irradiation. *Green Chemistry*, *6*(4), 219–221. https://doi.org/10.1039/b401620c.

Katritzky, A. R., Ramsden, C. A., Joule, J. A., & Zhdankin, V. V. (2010). Synthesis of monocyclic rings with one heteroatom. In A. R. Katritzky, C. A. Ramsden, J. A. Joule, & V. V. Zhdankin (Eds.), *Handbook of Heterocyclic Chemistry* (3rd ed., pp. 652–703). Elsevier. https://doi.org/10.1016/B978-0-08-095843-9.00013-6.

Kauffman, G. B., Pentimalli, R., Doldi (1908–2001), S., & Hall, M. D. (2010). Michele Peyrone (1813–1883), discoverer of cisplatin. *Platinum Metals Review*, *54*(4), 250–256. https://doi.org/10.1595/147106710X534326.

Koch, R. A., Yoon, G. M., Aryal, U. K., Lail, K., Amirebrahimi, M., LaButti, K., Lipzen, A., Riley, R., Barry, K., Henrissat, B., Grigoriev, I. V., Herr, J. R., & Aime, M. C. (2021). Symbiotic nitrogen fixation in the reproductive structures of a basidiomycete fungus. *Current Biology*, *31*(17), 3905-3914.e6. https://doi.org/10.1016/j.cub.2021.06.033.

Kolen, M., Smith, W. A., & Mulder, F. M. (2021). Accelerating ^{1}H NMR detection of aqueous ammonia. *ACS Omega*, *6*(8), 5698–5704. https://doi.org/10.1021/acsomega.0c06130.

Lemay, P., & Oesper, R. E. (1946). Claude Louis Berthollet (1748-1822). *Journal of Chemical Education*, *23*(4), 158–165. https://doi.org/10.1021/ed023p158.

Li, J., Lai, S., Chen, D., Wu, R., Kobayashi, N., Deng, L., & Huang, H. (2021). A review on combustion characteristics of ammonia as a carbon-free fuel. *Frontiers in Energy Research*, *9*, 760356. https://doi.org/10.3389/fenrg.2021.760356.

Lin, W., Luo, H., Wu, J., Hung, T.-C., Cao, B., Liu, X., Yang, J., & Yang, P. (2022). A Review of the emerging risks of acute ammonia nitrogen toxicity to aquatic decapod crustaceans. *Water*, *15*(1), 27. https://doi.org/10.3390/w15010027.

Lombardo, P. A. (2017). Chemistry: The hidden war. *Nature*, *541*(7636), 154–155. https://doi.org/10.1038/541154a.

Ma, F., Guo, L., Li, Z., Zeng, X., Zheng, Z., Li, W., Zhao, F., & Yu, W. (2023). A Review of current advances in ammonia combustion from the fundamentals to applications in internal combustion engines. *Energies*, *16*(17), 6304. https://doi.org/10.3390/en16176304.

Ma, Q., Wang, J., Jin, S., & Shi, M. (2025). The impact of ignition and activation energy distribution on the combustion and emission characteristics of diesel-ammonia-natural gas engines. *Journal of the Energy Institute*, *118*, 101903. https://doi.org/10.1016/j.joei.2024.101903.

Ma, R., Li, K., Guo, Y., Zhang, B., Zhao, X., Linder, S., Guan, C., Chen, G., Gan, Y., & Meng, J. (2021). Mitigation potential of global ammonia emissions and related health impacts in the trade network. *Nature Communications*, *12*(1), 6308. https://doi.org/10.1038/s41467-021-25854-3.

McCrory, P. (2006). Smelling salts. *British Journal of Sports Medicine*, *40*(8), 659–660. https://doi.org/10.1136/bjsm.2006.029710.

Möller, A., & Meyer, G. (1992). The ammonium ion and the ammine ligand as internal reducing agents for platinum-group-metal complexes. *Thermochimica Acta*, *210*, 147–150. https://doi.org/10.1016/0040-6031(92)80284-4.

Mowafy, A. M., El-ftouh, E. A. A., Sdiek, M. Y., Abdelshafi, S. A., Sallam, A. A., Agha, M. S., & Abou Zeid, W. R. (2022). Nitrogen-fixing archaea and sstainable agriculture. In D. K. Maheshwari, R. Dobhal, & S. Dheeman (Eds.), *Microorganisms for Sustainability* (Vol. 36, pp. 115–126). Springer. https://doi.org/10.1007/978-981-19-4906-7_6.

NIST. (2025, January 8). *Ammonia*. Https://Webbook.Nist.Gov/Cgi/Cbook.Cgi?ID=C7664417&Mask=200#Mass-Spec..

Norman, R. O. C., & Coxon, J. M. (2017). *Principles of Organic Synthesis* (R. O. C. Norman & J. M. Coxon, Eds.; 3rd ed.). Routledge. https://doi.org/10.1201/9780203742068.

Oommen, C., & Jain, S. R. (1999). Ammonium nitrate: A promising rocket propellant oxidizer. *Journal of Hazardous Materials*, *67*(3), 253–281. https://doi.org/10.1016/S0304-3894(99)00039-4.

Ouellette, R. J., & Rawn, J. D. (2014). Amines and amides. In R. J. Ouellette & J. D. Rawn (Eds.), *Organic Chemistry: Structure, Mechanism, and Synthesis* (pp. 803–842). Elsevier. https://doi.org/10.1016/B978-0-12-800780-8.00023-1.

Our World in Data. (2025, January 8). *Ammonia*. Https://Ourworldindata.Org/Search?Q=ammonia..

Preuss, M., & Bechstedt, F. (2006). Vibrational spectra of ammonia, benzene, and benzene adsorbed on Si(001) by first principles calculations with periodic boundary conditions. *Physical Review B*, *73*(15), 155413. https://doi.org/10.1103/PhysRevB.73.155413.

Reti, L. (1965). How old is hydrochloric acid? *Chymia*, *10*, 11–23. https://doi.org/10.2307/27757245.

Ridgway, J., Appleton, J. D., & Levinson, A. A. (1990). Ammonium geochemistry in mineral exploration—a comparison of results from the American cordilleras and the southwest Pacific. *Applied Geochemistry*, *5*(4), 475–489. https://doi.org/10.1016/0883-2927(90)90022-W.

Ritchie, H. (2025, January 8). *How many people does synthetic fertilizer feed?* Https://Ourworldindata.Org/How-Many-People-Does-Synthetic-Fertilizer-Feed.

Schranck, J., & Tlili, A. (2018). Transition-metal-catalyzed monoarylation of ammonia. *ACS Catalysis*, *8*(1), 405–418. https://doi.org/10.1021/acscatal.7b03215.

Sheeran, J. (2011). Preventing hunger: Sustainability not aid. *Nature*, *479*(7374), 469–470. https://doi.org/10.1038/479469a.

Sheppard, D. (2017). Robert Le Rossignol, 1884–1976: Engineer of the ‘Haber’ process. *Notes and Records: The Royal Society Journal of the History of Science*, *71*(3), 263–296. https://doi.org/10.1098/rsnr.2016.0019.

Shimizu, S., Abe, N., Iguchi, A., Dohba, M., Sato, H., & Hirose, K. (1998). Synthesis of pyridine bases on zeolite catalyst. *Microporous and Mesoporous Materials*, *21*(4–6), 447–451. https://doi.org/10.1016/S1387-1811(98)00052-3.

Short, D., & Crook, M. (2022). *The Genocide-Ecocide Nexus* (D. Short & M. Crook, Eds.; 1st ed.). Routledge. https://doi.org/10.4324/9781003253983.

Statista. (2025a, January 8). *Production capacity of ammonia worldwide from 2018 to 2023, with a forecast for 2026 and 2030.* Https://Www.Statista.Com/Statistics/1065865/Ammonia-Production-Capacity-Globally/.

Statista. (2025b, January 8). *Production of ammonia worldwide from 2010 to 2023.* Https://Www.Statista.Com/Statistics/1266378/Global-Ammonia-Production/.

Stephens, H. W. (1997). *The Texas City Disaster, 1947* (H. W. Stephens, Ed.). University of Texas Press. https://doi.org/10.7560/777224.

Sutton, M. A., van Dijk, N., Levy, P. E., Jones, M. R., Leith, I. D., Sheppard, L. J., Leeson, S., Sim Tang, Y., Stephens, A., Braban, C. F., Dragosits, U., Howard, C. M., Vieno, M., Fowler, D., Corbett, P., Naikoo, M. I., Munzi, S., Ellis, C. J., Chatterjee, S., … Wolseley, P. A. (2020). Alkaline air: Changing perspectives on nitrogen and air pollution in an ammonia-rich world. *Philosophical Transactions of the Royal Society A: Mathematical, Physical and Engineering Sciences*, *378*(2183), 20190315. https://doi.org/10.1098/rsta.2019.0315.

The Oppau explosion. (1921). *Nature*, *108*(2713), 278–279. https://doi.org/10.1038/108278a0.

Tonelli, D., Rosa, L., Gabrielli, P., Parente, A., & Contino, F. (2024). Cost-competitive decentralized ammonia fertilizer production can increase food security. *Nature Food*, *5*(6), 469–479. https://doi.org/10.1038/s43016-024-00979-y.

Valera-Medina, A., Amer-Hatem, F., Azad, A. K., Dedoussi, I. C., de Joannon, M., Fernandes, R. X., Glarborg, P., Hashemi, H., He, X., Mashruk, S., McGowan, J., Mounaim-Rouselle, C., Ortiz-Prado, A., Ortiz-Valera, A., Rossetti, I., Shu, B., Yehia, M., Xiao, H., & Costa, M. (2021). Review on ammonia as a potential fuel: from synthesis to economics. *Energy & Fuels*, *35*(9), 6964–7029. https://doi.org/10.1021/acs.energyfuels.0c03685.

Wang, H., Lin, N., & Arzumanyan, M. (2024). The market for low-carbon-intensity ammonia. *Gases*, *4*(3), 224–235. https://doi.org/10.3390/gases4030013.

Wentrup, C. (2024). The transition from alchemical to modern chemical symbolism: from Bergman and Guiton de Morveau to Hassenfratz and Adet, Higgins, Richter, Dalton, and Berzelius. *ChemPlusChem*, *89*(7), e202400033. https://doi.org/10.1002/cplu.202400033.

Wiley, R. H., Quilico, A., Speroni, G., Behr, L. C., & McKee, R. L. (1962). *Chemistry of Heterocyclic Compounds* (R. H. Wiley, A. Quilico, G. Speroni, L. C. Behr, & R. L. McKee, Eds.; Vol. 17). Wiley. https://doi.org/10.1002/9780470186787.

Wilson, J. J., & Lippard, S. J. (2014). Synthetic methods for the preparation of platinum anticancer complexes. *Chemical Reviews*, *114*(8), 4470–4495. https://doi.org/10.1021/cr4004314.

World Health Organization. (2025, January 8). *Model List of Essential Medicines.* Https://List.Essentialmeds.Org/.

Yüzbaşıoğlu, A. E., Tatarhan, A. H., & Gezerman, A. O. (2021). Decarbonization in ammonia production, new technological methods in industrial scale ammonia production and critical evaluations. *Heliyon*, *7*(10), e08257. https://doi.org/10.1016/j.heliyon.2021.e08257.

Zarebska, A., Romero Nieto, D., Christensen, K. V., Fjerbæk Søtoft, L., & Norddahl, B. (2015). Ammonium fertilizers production from manure: A critical review. *Critical Reviews in Environmental Science and Technology*, *45*(14), 1469–1521. https://doi.org/10.1080/10643389.2014.955630.

Zhang, T.-X., Li, M.-R., Liu, C., Wang, S.-P., & Yan, Z.-G. (2023). A review of the toxic effects of ammonia on invertebrates in aquatic environments. *Environmental Pollution*, *336*, 122374. https://doi.org/10.1016/j.envpol.2023.122374.

Zhao, Y., Ding, Y., Li, W., Liu, C., Li, Y., Zhao, Z., Shan, Y., Li, F., Sun, L., & Li, F. (2023). Efficient urea electrosynthesis from carbon dioxide and nitrate via alternating Cu–W bimetallic C–N coupling sites. *Nature Communications*, *14*(1), 4491. https://doi.org/10.1038/s41467-023-40273-2.

Zimmerman, H. E. (2012). A mechanistic analysis of the Birch reduction. *Accounts of Chemical Research*, *45*(2), 164–170. https://doi.org/10.1021/ar2000698.

Index

P

R

S

T

V

W